Robert McAndrew

List of the British Marine Invertebrate Fauna

Robert McAndrew

List of the British Marine Invertebrate Fauna

ISBN/EAN: 9783337272753

Printed in Europe, USA, Canada, Australia, Japan

Cover: Foto ©berggeist007 / pixelio.de

More available books at **www.hansebooks.com**

LIST

OF THE

BRITISH MARINE INVERTEBRATE FAUNA.

FOR THE DREDGING COMMITTEE OF THE BRITISH ASSOCIATION.

LONDON:

TAYLOR AND FRANCIS, RED LION COURT, FLEET STREET.

1860.

NOTICE.

The following lists have been prepared in conformity with the desire of the Committee of the Natural History Section of the British Association for the Advancement of Science, who, at my suggestion, recommended the appointment of a general Dredging Committee, with a liberal grant of money for the carrying out of its objects. .

It is intended to place these lists in the hands of the local Dredging Committees and naturalists engaged in researches in the most important districts of the coasts of Great Britain and Ireland, with a request that they may be returned, with notes on the conditions under which each species of the particular district has been found, and memoranda of such additional species as may be obtained. By this means it is hoped to collect local lists of great interest, and materials for a more complete catalogue of the Invertebrate Fauna of the British Seas. In the preparation of the present lists, I have been assisted by Dr. Baird and Mr. S. Woodward, and other members of the Dredging Committee. The catalogue of Mollusca is taken from the work of Messrs. Forbes and Hanley; that of Crustacea has been obligingly furnished by Mr. Spence Bate; of Radiata by Mr. Stewart, of the Royal College of Surgeons; of Sponges by Dr. Bowerbank; of Rhizopoda by Messrs. Rupert Jones and Parker; and to Dr. J. E. Gray I am indebted for permission to extract the list of Anellides from an unpublished work by the late Dr. Johnston, of Berwick-upon-Tweed.

ROBERT McANDREW.

Isleworth House, Feb. 10, 1860.

⁎ The nomenclature and arrangement are taken (with a few slight modifications) from the " British Mollusca " of Messrs. Forbes and Hanley.

CEPHALOPODA.

Species.	Range.	Found living at	Ground.	Frequency.	Observations.
	fathoms.	fathoms.			
OCTOPUS, *Cuvier.*					
vulgaris, *Lam.*					
ELEDONE, *Leach.*					
octópodia, *Penn.*					
ROSSIA, *Owen.*					
Owenii, *Ball*					
macrosoma, *Delle Chiaje*					
SEPIOLA, *Leach.*					
Rondeletii, *Leach*					
Atlantica, *D'Orbigny*					
OMMASTREPHES, *D'Orbigny.*					
sagittatus, *Lam.*					
todarus, *Delle Chiaje*					
Eblanæ, *Ball*					
LOLIGO, *Lamarck.*					
vulgaris, *Lam.* ? (Forbesi, *Stp.*)					
media, *Linn.*					
marmoræ, *Verany* (media, *var.*)					
SEPIA, *Linnæus.*					
officinalis, *Linn.*					
elegans, *Bl.*					
biserialis, *De Montfort*					

GASTEROPODA.

ORDER PROSOBRANCHIATA.

MUREX, *Linnæus.*					
erinaceus, *Linn.*					
corallinus, *Scacchi*					
TROPHON, *Montfort.*					
clathratus, *Linn.*					
muricatus, *Mont.*					
Barviecnsis, *Johnston*					
FUSUS, *Lamarck.*					
gracilis, *Da Costa*					
propinquus, *Alder*					
Berniciensis, *King*					
antiquus, *Linn.*					
Norvegicus, *Chemn.*					
Turtoni, *Bean*					
BUCCINUM, *Linnæus.*					
undatum, *Linn.*					
Dalei, *J. Sowerby*					
Humphresianum, *Bennett*					
fusiforme, *Broderip*					

3

A 2

Species.	Range.	Found living at	Ground.	Frequency.	Observations.
	fathoms.	fathoms			
NASSA, *Lamarck.*					
reticulata, *Linn.*					
pygmæa, *Lamarck*					
incrassata, *Müller*					
PURPURA, *Lamarck.*					
lapillus, *Linn.*					
MANGELIA, *Leach.*					
attenuata, *Mont.*					
costata, *Pennant*					
brachystoma, *Philippi*					
gracilis, *Mont.*					
Leufroyi, *Michaud*					
linearis, *Mont.*					
nana, *Lovén*					
nebula, *Mont.*					
purpurea, *Mont.*					
rufa, *Mont.*					
septangularis, *Mont.*					
striolata, *Scacchi*					
teres, *Forbes*					
Trevelliana, *Turton*					
turricula, *Mont.*					
LACUNIS, *Risso.*					
minima, *Mont.*					
MARGINELLA, *Lamarck.*					
lævis, *Donovan*					
OVULA, *Lamarck.*					
patula, *Pennant*					
? acuminata, *Bruguière*					
CYPRÆA, *Linnæus.*					
Europæa, *Mont.*					
NATICA, *Lamarck.*					
monilifera, *Lamarck*					
nitida, *Donovan*					
sordida, *Philippi*					
Montagui, *Forbes*					
helicoides, *Johnston*					
pusilla, *Say*					
Kingii, *Forbes*					
LAMELLARIA, *Montagu.*					
perspicua, *Linn.*					
tentaculata, *Mont.*					
VELUTINA, *Fleming.*					
flexilis, *Mont.*					
lævigata, *Linn.*					
? OTINA, *Gray.*					
otis, *Turton*					
TRICHOTROPIS, *Broderip.*					
borealis, *Brod.*					

4

Species.	Range.	Found living at	Ground.	Fre-quency.	Observations.
	fathoms	fathoms.			
CERITHIOPSIS, *Forbes & Hanley.*					
tubercularis, *Mont.*					
ODOSTOMIA, *Fleming.*					
acuta, *Jeffreys*					
alba, *Jeffreys*					
conoidea, *Brocchi*					
conspicua, *Alder*					
cylindrica, *Alder*					
decussata, *Mont.*					
dolioliformis, *Jeffreys*					
dubia, *Jeffreys*					
eulimoides, *Hanley*					
excavata, *Philippi*					
glabrata, *Mühlfeldt*					
Gulsonæ, *Clark*					
insculpta, *Mont.*					
interstincta, *Mont.*					
minuta, *Jeffreys*					
nitida, *Alder*					
obliqua, *Alder*					
pallida, *Mont.*					
plicata, *Mont.*					
rissoides, *Hanley*					
spiralis, *Mont.*					
striolata, *Alder*					
truncatula, *Jeffreys*					
unidentata, *Mont.*					
Warrenii, *Thompson*					
EULIMELLA, *Forbes.*					
acicula, *Philippi*					
affinis, *Philippi*					
clavula, *Lovén*					
Scillæ, *Scacchi*					
CHEMNITZIA, *D'Orbigny.*					
clathrata, *Jeffreys*					
elegantissima, *Mont.*					
fenestrata, *Jeffreys*					
formosa, *Jeffreys*					
fulvocincta, *Thomps.*					
indistincta, *Mont.*					
rufa, *Philippi*					
rufescens, *Forbes*					
scalaris, *Philippi*					
eximia, *Jeffreys*					
EULIMA, *Risso.*					
polita, *Linn.*					
distorta, *Deshayes*					
subulata, *Donovan*					
bilineata, *Alder*					

5

Species.	Range.	Found living at	Ground.	Frequency.	Observations.
	fathoms.	fathoms.			
STYLINA, *Fleming.*					
Turtoni. *Broderip*					
CERITHIUM, *Bruguière.*					
metula, *Lovén*					
reticulatum. *Da Costa*					
adversum, *Mont.* ....,....					
APORRHAIS. *Aldrovandus.*					
pes-carbonis, *Brongniart*					
pes-pelecani, *Linn.*					
TURRITELLA, *Lamarck.*					
communis. *Risso*					
ACLIS, *Lovén.*					
ascaris, *Turton*					
supranitida, *S. Wood*					
? unica, *Mont.*					
nitidissima. *Mont.*					
CÆCUM, *Fleming.*					
trachea, *Mont.*					
glabrum, *Mont.*					
SCALARIA, *Lamarck.*					
Turtoni, *Turton*					
communis, *Lamarck*					
clathratula, *Mont.*					
Grœnlandica, *Chemnitz*					
Trevelyana, *Leach*					
SKENEA, *Fleming.*					
? costulata, *Müller*					
? lævis, *Philippi*					
planorbis, *Fabr.*					
? nitidissima, *Adams*					
? rota, *Forbes*					
TRUNCATELLA, *Risso.*					
Montagui, *Lowe*					
JEFFREYSIA. *Alder.*					
opalina, *Jeffreys*					
diaphana, *Alder*					
globularis, *Jeffreys*					
RISSOA, *Frémenville.*					
abyssicola. *Forbes*					
anatina, *Drap.*					
Beanii, *Hanley*					
calathus, *Forbes*					
cingillus. *Mont.*					
costata, *Adams*					
costulata, *Risso*					
crenulata, *Michaud*					
fulgida. *Adams*					
inconspicua, *Alder*					
labiosa, *Mont.*					

6

GASTEROPODA.

Species.	Range.	Found living at	Ground.	Fre- quency.	Observations.
	fathoms	fathoms			
RISSOA. *Frémenville.*					
lactea, *Michaud*					
littorea, *Delle Chiaje*					
muriatica, *Lam.*					
parva, *Da Costa*					
proxima, *Alder*					
pulcherrima, *Jeffreys*					
punctura, *Mont.*					
rubra, *Adams*					
rufilabrum, *Alder*					
sculpta, *Philippi*					
semistriata, *Mont.*					
soluta, *Philippi*					
striata, *Mont.*					
striatula, *Mont.*					
ulvæ, *Pennant*					
ventrosa, *Mont.*					
vitrea, *Mont.*					
Zetlandica, *Mont.*					
ASSIMINEA. *Leach.*					
Grayana, *Leach*					
LACUNA, *Turton.*					
crassior, *Mont.*					
vincta, *Mont.*					
puteolus, *Turton*					
pallidula, *Da Costa*					
LITORINA. *Férussac.*					
fabalis, *Turton*					
litoralis, *Linn.*					
litorea, *Linn.*				•	
neritoides, *Linn.*					
tenebrosa, *Mont.*					
palliata, *Say*					
patula, *Jeffreys*					
rudis, *Donovan*					
saxatilis, *Johnston*					
ADEORBIS, *S. Wood.*					
subcarinata, *Mont.*					
divisa, *Fleming*					
TROCHUS, *Linn.*					
alabastrum, *Beck*					
cinerarius, *Linn.*					
conulus, *Linn.*					
exiguus, *Pulteney*					
granulatus, *Born*					
crassus, *Pulteney* (lineatus, *Da Costa*)					
magus, *Linn.*					
millegranus, *Philippi*					

7

Species.	Range.	Found living at	Ground	Fre-quency.	Observations
	fathoms.	fathoms.			
TROCHUS, *Linn.*					
Montagui, *Gray*					
striatus, *Linn.*					
tumidus, *Mont.*					
umbilicatus, *Mont.*					
lineatus, *Da Costa*					
zizyphinus, *Linn.*					
MARGARITA, *Leach.*					
undulata, *Sowerby*					
helicina, *Fabr*					
pusilla, *Jeffreys*					
Cutleriana, *Clark*					
PHASIANELLA, *Lamarck.*					
pullus, *Linn.*					
IANTHINA, *Lamarck.*					
exigua, *Lamarck*					
communis, *Lamarck*					
pallida, *Harvey*					
SCISSURELLA, *D'Orbigny.*					
crispata, *Fleming*					
HALIOTIS, *Linn.*					
tuberculata, *Linn.*					
EMARGINULA, *Lamarck.*					
reticulata, *J. Sow.*					
rosea, *Bell*					
crassa, *J. Sow.*					
PUNCTURELLA, *Lowe.*					
Noachina, *Linn.*					
FISSURELLA, *Lamarck.*					
reticulata, *Donovan*					
PILEOPSIS, *Lamarck.*					
Hungaricus, *Linn.*					
CALYPTRÆA, *Lamarck.*					
Sinensis, *Linn.*					
ACMÆA, *Eschscholtz.*					
testudinalis, *Müller*					
virginea, *Müller*					
PATELLA, *Linnæus.*					
vulgata, *Linn.*					
athletica, *Bean*					
pellucida, *Linn*					
lævis, *Pennant*					
PILIDIUM, *Forbes.*					
fulvum, *Müller*					
PROPILIDIUM, *Forbes.* (LEPETA, *Gr.*)					
ancyloide, *Forbes*					
DENTALIUM, *Linnæus.*					
entale, *Linn.*					
Tarentinum, *Lam.*					

8

GASTEROPODA.

Species.	Range.	Found living at	Ground	Fre-quency.	Observations.
	fathoms.	fathoms.			
CHITON, *Linnæus*.					
fascicularis, *Linn.*					
discrepans, *Brown*					
Hanleyi, *Bean*					
ruber, *Linn.*					
cinereus, *Linn*					
albus, *Linn.*					
asellus, *Chemn.*					
cancellatus, *Sow.*					
lævis, *Pennant*					
marmoreus, *O. Fabr.*					

ORDER OPISTHOBRANCHIATA.

Species.					
TORNATELLA, *Lamarck.*					
fasciata, *Linn.*					
BULLA, *Lamarck.*					
hydatis, *Linn.*					
Cranchii, *Leach*					
AKERA, *Müller.*					
bullata, *Linn.*					
CYLICHNA, *Lovén.*					
cylindracea, *Pennant*					
conulus, *Desh.*					
mamillata, *Philippi*					
nitidula, *Lovén*					
obtusa, *Mont.*					
strigella, *Lovén*					
truncata, *Adams*					
umbilicata, *Mont.*					
AMPHISPHYRA, *Lovén.*					
hyalina, *Turton*					
SCAPHANDER, *Montfort.*					
lignarius, *Linn.*					
BULLÆA, *Lamarck.*					
aperta, *Linn.*					
quadrata, *S. Wood*					
scabra, *Müller*					
catena, *Mont.*					
punctata, *Clark*					
pruinosa, *Clark*					
APLYSIA, *Gmelin.*					
hybrida, *Sow.*					
PLEUROBRANCHUS, *Cuvier.*					
plumula, *Mont.*					
membranaceus, *Mont.*					
DIPHYLLIDIA, *Cuvier.*					
lineata, *Otto*					

D

GASTEROPODA.

ORDER NUDIBRANCHIATA.

(From the Monograph of Messrs. Alder and Hancock, 1855.)

Species.	Range.	Found living at	Ground.	Frequency.	Observations.
	fathoms.	fathoms			
Doris, *Linn.*					
tuberculata, *Cuv.*					
flammea, *A. & H.*					
Zetlandica, *A. & H.*					
millegrana, *A. & H.*					
Johnstoni, *A. & H.* . . .					
planata, *A. & H.*					
coccinea, *Forbes*					
repanda, *A. & H.*					
aspera, *A. & H.*					
proxima, *A. & H.*					
muricata, *Müll.* . . .					
Ulidiana, *Thomps.*					
diaphana, *A. & H.*					
oblonga, *A. & H.*					
bilamellata, *L.*					
depressa, *A. & H.*					
inconspicua, *A. & H.*					
pusilla, *A. & H.*					
sparsa, *A. & H.*					
pilosa, *Müll.*					
subquadrata, *A. & H.*					
Goniodoris, *Forbes.*					
nodosa, *Mont.*					
castanea, *A. & H.*					
Triopa, *Johnston.*					
claviger, *Müll.* . . .					
Ægirus, *Lovén.*					
punctilucens, *D'Orb.*					
Thecacera, *Fleming.*					
pennigera, *Mont.*					
virescens, *A. & H.* . . .					
capituta, *A. & H.*					
Polycera, *Cuvier.*					
quadrilineata, *Müll.* . . .					
ocellata, *A. & H.*					
Lessonii. *D'Orb.*					
Ancula, *Lovén.*					
cristata, *Alder* . .					
Idalia, *Leuckart.*					
elegans, *Leuck.*					
Leachii, *A. & H.*					
aspersa, *A. & H.* . .					

10

Species.	Range.	Found living at	Ground.	Frequency.	Observations
	fathoms.	fathoms.			
IDALIA, *Leuckart.*					
inæqualis, *Forbes*					
pulchella, *A. & H.*					
quadricornis, *Mont.*					
TRITONIA, *Cuvier.*					
Hombergii, *Cuv.*					
alba, *A. & H.*					
plebeia, *Johnston*					
lineata, *A. & H.*					
SCYLLÆA, *Linn.*					
pelagica, *Linn.*					
LOMANOTUS, *Verany.*					
marmoratus, *A. & H.*					
flavidus, *A. & H.*					
DENDRONOTUS, *A. & H.*					
arborescens, *Müll.*					
DOTO, *Oken.*					
fragilis, *Forbes*					
pinnatifida, *Mont.*					
coronata, *Müll.*					
ÆOLIS, *Cuvier.*					
papillosa, *Linn.*					
glauca, *A. & H.*					
Alderi, *Cocks*					
coronata, *Forbes*					
Drummondi, *Thomps.*					
punctata, *A. & H.*					
elegans, *A. & H.*					
rufibranchialis, *Johnst.*					
lineata, *Lov.*					
smaragdina, *A. & H.*					
gracilis, *A. & H.*					
pellucida, *A. & H.*					
Landsburgii, *A. & H.*					
alba, *A. & H.*					
carnea, *A. & H.*					
glaucoides, *A. & H.*					
Peachii, *A. & H.*					
nana, *A. & H.*					
stipata, *A. & H.*					
angulata, *A. & H.*					
inornata, *A. & H.*					
concinna, *A. & H.*					
olivacea, *A. & H.*					
aurantiaca, *A. & H.*					
pustulata, *A. & H.*					
Couchii, *Cocks*					
amœna, *A. & H.*					
Northumbrica, *A. & H.*					

Species.	Range.	Found living at	Ground.	Frequency.	Observations.
	fathoms	fathoms.			
Æolis, *Cuvier.*					
arenicola, *Forbes*					
Glottensis, *A. & H.*					
viridis, *Forbes*					
purpurascens, *Flem.*					
cingulata, *A. & H.*					
vittata, *A. & H.*					
cærulea, *Mont.*					
picta, *A. & H.*					
tricolor, *Forbes*					
amothystina, *A. & H.*					
Farrani, *A. & H.*					
exigua, *A. & H.*					
despecta, *Johnst.*					
Embletonia, *A. & H.*					
pulchra, *A. & H.*					
minuta, *F. & G.*					
pallida, *A. & H.*					
Fiona, *A. & H.*					
nobilis, *A. & H.*					
Hermæa, *Lovén.*					
bifida, *Mont.*					
dendritica, *A. & H.*					
Alderia, *Allman.*					
modesta, *Lovén*					
Proctonotus, *A. & H.*					
mucroniferus, *A. & H.*					
Antiopa, *A. & H.*					
cristata, *Del. Ch.*					
hyalina, *A. & H.*					

PTEROPODA.

Hyalæa, *Lamarck.*					
trispinosa, *Lesueur*					
Spirialis, *Eydoux & Souleyet.*					
Flemingii, *Forbes*					
Jeffreysii, *Forbes*					
MacAndrei, *Forbes*					
Clio, *Müller.*					
borealis, *Linn.*					

12

LAMELLIBRANCHIATA.

Species.	Range.	Found living at	Ground.	Frequency.	Observations.
	fathoms.	fathoms.			
OSTREA, *Linnæus.*					
edulis, *Linn.*					
ANOMIA, *Linnæus.*					
aculeata, *Müller*					
ephippium, *Linn.*					
striata, *Lovén*					
patelliformis, *Linn.*					
PECTEN, *O. F. Müller.*					
Danicus, *Chemnitz*					
maximus, *Linn.*					
niveus, *Macgillivray*					
opercularis, *Linn.*					
pusio, *Pennant*					
similis, *Laskey*					
tigrinus, *Müller*					
varius, *Linn.*					
striatus, *Müller*					
furtivus, *Lovén*					
LIMA, *Bruguière.*					
hians, *Gmelia*					
Loscombii, *Sowerby*					
subauriculata, *Mont.*					
AVICULA, *Bruguière.*				•	
Tarentina, *Lam.*					
PINNA, *Linnæus.*					
pectinata, *Linn.*					
MYTILUS, *Linnæus.*					
edulis, *Lian.*					
MODIOLA, *Lamarck.*					
barbata, *Linn.*					
modiolus, *Lian.*					
phascolina, *Philippi*					
tulipa, *Lam.*		•			
CRENELLA, *Brown.*					
costulata, *Risso*					
decussata, *Mont.*					
discors, *Linn.*					
nigra, *Gray*					
marmorata, *Forbes*					
rhombea, *Berkeley*					
AUCA, *Linnæus.*					
lactea, *Lian.*					
raridentata, *S. Wood*					
tetragona, *Poli*					
PECTUNCULUS, *Lamarck.*					
glycimeris, *Linn.*					

13

Species.	Range.	Found living at	Ground	Fre-quency.	Observations.
	fathoms.	fathoms.			
NUCULA, *Lamarck.*					
decussata, *Sow.*					
nitida, *Sow.*					
nucleus, *Linn.*					
radiata, *Hanley*					
tenuis, *Mont.*					
LEDA, *Schumacher.*					
caudata, *Don.*					
pygmæa, *Münster*					
CARDIUM, *Linnæus.*					
aculeatum, *Linn.*					
echinatum, *Linn.*					
edule, *Linn.*					
fasciatum, *Mont.*					
nodosum, *Turton*					
Norvegicum, *Speng.*					
pygmæum, *Don.* ...					
rusticum, *Linn.* ...					
Succicum, *Reeve* ...					
LUCINA, *Bruguière.*					
borealis, *Linn.*					
divaricata, *Linn.*					
ferruginosa, *Forbes*					
flexuosa, *Mont.*...					
leucoma, *Turton*					
spinifera, *Mont.*					
DIPLODONTA, *Bronn.*					
rotundata, *Mont.*					
KELLIA, *Turton.*					
suborbicularis, *Mont.*					
rubra, *Mont.*					
TURTONIA, *Hanley.*					
minuta, *O. Fabr.*					
MONTACUTA, *Turton.*					
bidentata, *Mont.*					
ferruginosa, *Mont.*					
substriata, *Mont.*					
LEPTON, *Turton.*					
Clarkiæ, *Clark*					
nitidum, *Turton*					
squamosum, *Mont.*					
GALEOMMA, *Turton.*					
Turtoni, *Sow.*					
CYPRINA, *Lamarck.*					
Islandica, *Linn.*					
CIRCE, *Schumacher.*					
minima, *Mont.*					
ASTARTE, *Sowerby.*					
arctica, *Gray*					

14

LAMELLIBRANCHIATA.

Species.	Range.	Found living at	Ground.	Fre-quency.	Observations.
	fathoms	fathoms.			
Astarte, *Sowerby.*					
compressa, *Mont.*					
crebricostata, *Forbes*					
elliptica, *Brown*					
sulcata, *Da Costa*					
triangularis, *Mont.*					
Isocardia, *Lamarck.*					
cor, *Linn.*					
Venus, *Linnæus.*					
casina, *Linn.*					
fasciata, *Don.*					
ovata, *Pennant*					
striatula, *Don.*					
verrucosa, *Linn.*					
Cytherea, *Lamarck.*					
chione, *Linn.*					
Artemis, *Poli.*					
exoleta, *Linn.*					
lincta, *Pult.*					
Lucinopsis, *Forbes.*					
undata, *Penn.*					
Tapes, *Mühlfeldt.*					
aurea, *Gmelin*					
decussata, *Linn.*					
pullastra, *Wood*					
virginea, *Linn.*					
Venerupis, *Lamarck.*					
irus, *Linn.*					
Petricola, *Lamarck.*					
lithophaga, *Retzius*					
Mactra, *Linnæus.*					
elliptica, *Brown*					
helvacea, *Chemnitz*					
solida, *Linn.*					
stultorum, *Linn.*					
subtruncata, *Da Costa*					
truncata, *Mont.*					
Lutraria, *Lamarck.*					
elliptica, *Linn.*					
oblonga, *Chemn.*					
Tellina, *Linnæus.*					
balaustina, *Linn.*					
crassa, *Penn.*					
donacina, *Linn.*					
fabula, *Gronov.*					
incarnata, *Linn.*					
proxima, *Brown*					
pygmæa, *Philippi*					
solidula, *Pult.*					

15

Species.	Range.	Found living at	Ground.	Fre-quency	Observations.
	fathoms.	fathoms.			
TELLINA, *Linnæus.*					
tenuis, *Da Costa*					
GASTRANA, *Sch.* (DIODONTA, *F.&H.*)					
fragilis, *Linn.*					
PSAMMOBIA, *Lamarck.*					
costulata, *Turt.*					
Ferroensis, *Chemn.*					
tellinella, *Lam.*					
vespertina, *Chemn.*					
SYNDOSMYA, *Recluz.*					
alba, *Wood*					
intermedia, *Thomps.*					
prismatica, *Mont.*					
tenuis, *Mont.*					
SCROBICULARIA, *Schumacher.*					
piperata, *Gmelin*					
ERVILIA, *Turton.*					
castanea, *Mont.*					
DONAX, *Linnæus.*					
anatinus, *Lam.*					
politus, *Poli*					
SOLEN, *Linnæus.*					
ensis, *Linn.*					
marginatus, *Pult.*					
pellucidus, *Penn.*					
, siliqua, *Linn.*					
CERATISOLEN, *Forbes.*					
legumen, *Linn.*					
SOLECURTUS, *Blainville.*					
candidus, *Renieri*					
coarctatus, *Gmelin*					
MYA, *Linnæus.*					
arenaria, *Linn.*					
truncata, *Linn.*					
CORBULA, *Bruguière.*					
nucleus, *Lam.*					
ovata, *Forbes*					
rosea, *Brown*					
SPHENIA, *Turton.*					
Binghami, *Turton*					
NEÆRA, *Gray.*					
abbreviata, *Forbes*					
costellata, *Desh.*					
cuspidata, *Olivi*					
POROMYA, *Forbes.* (= THETIS, *Sby.*)					
granulata, *Nyst*					
PANOPÆA, *Menard de la Groye.*					
Norvegica, *Speny.*					

16

LAMELLIBRANCHIATA.

Species.	Range.	Found living at	Ground.	Fre-quency.	Observations.
SAXICAVA, *Bellevue.*	fathoms.	fathoms.			
arctica, *Linn.*					
rugosa, *Linn.*					
COCHLODESMA, *Leach.* (= PERIPLOMA, *Sch.*)					
prætenue, *Pult.*					
THRACIA, *Leach.*					
convexa, *Wood*					
distorta, *Mont.*					
phascolina, *Lam.*					
pubescens, *Pult.*					
villosinscula, *Macgill.*					
LYONSIA, *Turton.*					
Norvegica, *Chemn.*					
PANDORA, *Bruguière.*					
obtusa, *Leach*					
rostrata, *Lam.*					
GASTROCHÆNA. *Spengler.*					
modiolina, *Lam.*					
PHOLAS, *Linnæus.*					
candida, *Linn.*					
crispata, *Linn.*					
dactylus, *Linn.*					
parva, *Penn.*					
striata, *Linn.*					
PHOLADIDEA, *Turton.*					
lamellata, *Turton*					
papyracea, *Solander.*					
XYLOPHAGA, *Turton.*					
dorsalis, *Turton*					
TEREDO, *Adanson.*					
bipennata, *Turton*					
malleolus, *Turton*					
megotara, *Hanley*					
navalis, *Linn.*					
Norvegica, *Speng.*					
palmulata, *Lam.*					

BRACHIOPODA.

Species.	Range.	Found living at	Ground.	Fre-quency.	Observations.
CRANIA, *Retzius.*					
anomala, *Müller*					
RHYNCHONELLA, *Fischer.*					
psittacea, *Chemn.*					
TEREBRATULA, *Bruguière.*					
caput serpentis, *Linn.*					
cranium, *Müller*					
capsula, *Jeffreys*					
ARGIOPE, *Deslongchamps*					
cistellula, *Searles Wood*					

17   c

TUNICATA.

Species.	Range.	Found living at	Ground.	Frequency.	Observations.
	fathoms.	fathoms.			
APLIDIUM, *Savigny.*					
ficus, *Linn.*					
fallax, *Johnst.*					
nutans, *Johnst.*					
SIDNYUM. *Savigny.*					
turbinatum, *Savig.*					
POLYCLINUM, *Savigny.*					
aurantium, *M.-Edw.*					
AMOUROUCIUM, *M.-Edw.*					
proliferum, *M.-Edw.*					
Nordmanni, *M.-Edw.*					
Argus, *M.-Edw.*					
LEPTOCLINUM, *M.-Edw.*					
maculosum, *M.-Edw.*					
asperum, *M.-Edw.*					
aureum, *M.-Edw*					
gelatinosum, *M.-Edw.*					
Listerianum, *M.-Edw.*					
punctatum, *Forbes*					
DISTOMA, *Gaertner.*					
rubrum, *Savig.*					
variolosum, *Gaertner*					
BOTRYLLUS, *Gaertner.*					
Schlosseri, *Pallas*					
polycyclus, *Savig.*					
gemmeus, *Savig.*					
violaceus, *M.-Edw.*					
smaragdus, *M.-Edw.*					
virescens, *A. & H.*					
bivittatus, *M.-Edw.*					
rubens, *A. & H.*					
castaneus, *A. & H.*					
BOTRYLLOIDES, *M.-Edw.*					
Leachii, *Savig.*					
ramulosa, *A. & H.*					
albicans, *M.-Edw.*					
radiata, *A. & H.*					
rotifera, *M.-Edw.*					
rubra, *M.-Edw.*					
CLAVELINA, *Savigny.*					
lepadiformis, *O. F. Müller*					
PEROPHORA. *Wiegmann.*					
Listeri, *Wiegm.*					
SYNTETHYS. *Forbes & Goodsir.*					
Hebridicus, *F. & G.*					

18

Species.	Range.	Found living at	Ground.	Frequency.	Observations.
	fathoms.	fathoms.			
ASCIDIA, *Baster.*					
intestinalis, *Linn.*					
canina, *O. F. Müll.*					
venosa, *O. F. Müll.*					
mentula, *O. F. Müll.*					
arachnoïdea, *E. Forbes*					
scabra, *O. F. Müll.*					
virginea, *O. F. Müll.*					
parallelogramma, *O. F. Müll.* ...					
prunum, *Müll.?*					
orbicularis, *Müll.*					
depressa, *A. & H.*				·
aspersa, *Müll.*					
vitrea, *Van Beneden*					
conchilega, *O. F. Müll.*					
echinata, *Linn.*					
sordida, *A. & H.*					
albida, *A. & H.*					
elliptica, *A. & H.*					
pellucida, *A. & H.*					
MOLGULA, *E. Forbes.*					
oculata, *E. Forbes*					
arenosa, *A. & H.*					
CYNTHIA, *Savigny.*					
microcosmus, *Savig.*					
claudicans, *Savig.*					
tuberosa, *Macgillivray*					
quadrangularis, *E. Forbes*					
informis, *E. Forbes*					
tessellata, *E. Forbes*					
limacina, *E. Forbes*					
morus, *E. Forbes*					
rustica, *Linn.*					
grossularia, *Van Beneden*					
ampulla, *Brug.*					
mamillaris, *Pallas*					
aggregata, *Rathke*					
coriacea, *A. & H.*					
PELONAIA, *Forbes & Goodsir.*					
corrugata, *Forbes & Hanl.*					
glabra, *Forbes & Hanl.*					
SALPA, *Chamisso.*					
runcinata, *Cham.*					
APPENDICULARIA, *Chamisso.*					
sp.					

CRUSTACEA.

BRACHYURA.

Species.	Range.	Found living at	Ground.	Frequency.	Observations.
	fathoms.	fathoms.			
STENORHYNCHUS, *Lamarck*.					
phalangium, *Pennant*					
tenuirostris, *Leach*					
ACHÆUS, *Leach*.					
Cranchii. *Leach*					
INACHUS, *Fabr*.					
Dorsettensis, *Penn*.					
dorhynchus, *Leach*					
leptochirus, *Leach*					
PISA, *Leach* (AUCTORSIS, *Lam*.).					
tetraodon, *Leach*					
Gibbsii, *Leach* (lanata, *Lam*.) ..					
HYAS, *Leach*.					
araneus, *Fabr*.					
coarctatus, *Leach*					
MAIA, *Lam*.					
squinado, *Herbst*					
EURYNOME, *Leach*.					
aspera, *Leach*					
XANTHO, *Leach*.					
florida, *Leach*					
rivulosa, *Edw*.					
tuberculata, *Couch*					
CANCER, *Linn*.					
pagurus, *Linn*.					
PILUMNUS, *Leach*.					
hirtellus, *Leach*					
PIRIMELA, *Leach*.					
denticulata, *Mont*.					
CARCINUS, *Leach*.					
mœnas, *Linn*.					
PORTUMNUS, *Leach*.					
variegatus, *Leach* (latipes, *Penn*.)					
PORTUNUS, *Leach*.					
puber, *Linn*.					
corrugatus, *Leach*					
arcuatus, *Leach*					
depurator, *Leach*					
marmoreus, *Leach*					
holsatus, *Fabr*.					
pusillus, *Leach*					
longipes, *Risso*					
plicatus, *Risso*					
carcinoides, *Kin*.					
POLYBIUS, *Leach*.					
Henslowii, *Leach*					

CRUSTACEA.

Species.	Range.	Found living at	Ground.	Frequency.	Observations.
	fathoms.	fathoms.			
PINNOTHERES, *Latr.*					
pisum, *Penn.*					
veterum, *Bosc*					
GONOPLAX, *Leach.*					
angulata, *Leach*					
PLANES, *Leach.*					
Linnæana, *Leach*					
EBALIA, *Leach.*					
Pennantii (tuberosa, *Penn.*) ..					
Bryerii, *Leach* (tumefacta, *Mont.*)					
Cranchii, *Leach*					
ATELECYCLUS, *Leach.*					
heterodon, *Leach* (septomden-tatus, *Mont.*)					
CORYSTES, *Leach.*					
Cassivelaunus, *Leach*					
THIA, *Leach.*					
polita, *Leach*					

ANOMOURA.

DROMIA, *Edw.*					
vulgaris, *Edw.*					
LITHODES, *Latr.*					
Maia, *Leach*					
PAGURUS, *Fabr.*					
Bernhardus, *Linn.*					
Prideauxii, *Leach*					
Cuanensis, *Thompson*					
Ulidianus, *Thompson*					
Hyndmanni, *Thompson*					
lævis, *Thompson*					
Forbesii, *Bell*					
Thompsoni, *Bell*					
fasciatus, *Bell*					
Dillwynii, *Spence Bate*					
PORCELLANA, *Lamarck.*					
platycheles, *Penn.*					
longicornis, *Penn.*					
GALATHEA, *Fabr.*					
squamifera, *Leach*					
dispersa, *Spence Bate*					
strigosa, *Fabr.*					
nexa, *Emb.*					
Andrewsii, *Kinahan*					
MUNIDA, *Leach.*					
Bamfica, *Penn.* (Rondeletii, *Bell*)					

21

CRUSTACEA.

MACROURA.

Species.	Range.	Found living at	Ground.	Frequency.	Observations
	fathoms.	fathoms.			
SCYLLARUS, *Fabr.*					
arctus, *Linn.*					
PALINURUS, *Fabr.*					
Homarus, *Linn.*					
CALLIANASSA, *Leach.*					
subterranea, *Leach*					
GEBIA, *Leach.*					
stellata, *Mont.*					
deltura, *Leach*					
AXIUS, *Leach.*					
stirhynchus, *Leach*					
CALOCARIS, *Bell.*					
Macandreæ, *Bell*					
ASTACUS, *Fabr.*					
gammarus (*L.*) (marinus. *Fabr.*					
vulgaris, *Edw.*)					
NEPHROPS, *Leach.*					
Norvegicus, *Linn.*					
CRANGON, *Fabr.*					
vulgaris, *Fabr.*					
fasciatus, *Risso*					
spinosus, *Leach*					
sculptus, *Bell*					
trispinosus, *Hailstone*					
bispinosus, *Westw., Kinahan* ..					
Allmanni, *Kin.*					
Pattersonii, *Kin.*					
ALPHEUS, *Fabr.*					
ruber, *Edw.*					
affinis, *Guise*					
AUTONOMEA, *Risso.*					
Olivii, *Risso*					
NIKA, *Risso.*					
edulis, *Risso*					
Couchii, *Bell*					
ATHANAS, *Leach.*					
nitescens, *Mont., Leach*					
HIPPOLYTE, *Leach*					
spinus, *Sowerby*					
varians, *Leach*					
Cranchii, *Leach*					
Thompsoni, *Bell*	•				
Prideauxiana, *Leach*					
Gordoni, *Spence Bate*					
fascigera, *Gosse*					

22

CRUSTACEA.

Species.	Range.	Found living at	Ground.	Frequency.	Observations.
	fathoms.	fathoms.			
HIPPOLYTE, *Leach.*					
Grayana, *Thompson*					
Mitchelli, *Thompson*					
Whitei, *Thompson*					
Yarrellii, *Thompson*					
Barleei, *Spence Bate*					
pandaliformis, *Bell*					
pusiola, *Kröyer*					
PANDALUS, *Leach.*					
Jeffreysii, *Spence Bate, Kinahan*					
annulicornis, *Leach*					
leptorhynchus, *Kin.*					
PALÆMON, *Fabr.*					
serratus, *Penn.*					
squilla, *Fabr.*					
Leachii, *Bell*					
varians, *Leach*					
PASIPHLEA, *Savigny.*					
sivado, *Risso*					
PENÆUS, *Fabr.*					
caramote, *Rioss.*					

STOMAPODA.

MYSIS, *Latr.*					
chamæleon, *V. Thompson*					
vulgaris, *V. Thompson*					
Griffithsiæ, *Bell*					
Lamoruæ, *Couch*					
productus, *Gosse*					
Oberon, *Couch*					
THYSANOPODA, *Edw.*					
Couchii, *Bell*					
MACROMYSIS, *White* (THEMISTO, *Goodsir, Bell*).					
longispinosus, *Goodsir*					
brevispinosus, *Goodsir*					
CYNTHILIA, *White* (CYNTHIA, *V. Thomps., Bell*).					
Flemingii, *Goodsir*					
CUMA, *Edwards.*					
scorpioides, *Mont.*					
unguiculata, *Spence Bate*					
VAUNTHOMSONIA, *Spence Bate.*					
Edwardsii, *Kröyer*					
cristata. *Spence Bate*					

23

Species.	Range.	Found living at	Ground.	Fre-quency.	Observations.
	fathoms.	fathoms.			
DIASTYLIS, *Say* (ALAUNA, *Goodsir, Bell*).					
Rathkii, *Kr.* (rostrata, *Goodsir, Bell*)					
ECDORA, *Spence Bate.*					
truncatula, *Spence Bate*					
IPHITHOË, *Spence Bate* (HALIA, *Spence Bate, White*).					
trispinosa, *Goodsir*					
BODOTRIA, *Goodsir.*					
arenosa, *Goodsir*					
CYRIANASSA, *Spence Bate* (VENILIA, *Spence Bate, White*).					
gracilis, *Spence Bate*					
longicornis, *Spence Bate*					
SQUILLA, *Fabr.*					
Desmarestii, *Risso*					
mantis, *Rondelet*					
PHYLLOSOMA. *Leach.*					
Cranchii, *Leach*					

AMPHIPODA NORMALIA.

Species.					
TALITRUS, *Latr.*					
locusta, *Auct.*					
ORCHESTIA, *Leach.*					
littorea, *Mont.*					
Deshayesii, *Savig.*					
Mediterranea, *Costa* (lævis, *S. Bate* ; littorea, var., *White*)..					
ALLORCHESTES, *Dana.*					
Nilssonii, *Kröyer* (Danai, *Spence Bate*)					
imbricatus, *Spence Bate*					
NICEA, *Nicolet* (GALANTHIS, *Spence Bate*).					
Lubbockiana, *Spence Bate*					
MONTAGUA, *Spence Bate.*					
monoculoides, *Montagu* (Typhis monoculoides, *White, Gosse*)..					
marina, *Spence Bate*					
Alderii, *Spence Bate*					
pollexiana, *Spence Bate*			•		
DANAIA, *Spence Bate.*					
dubia, *Spence Bate*					
LYSIANASSA. *M.-Edw.*					
Costæ, *M.-Edw.*					

24

Species.	Range.	Found living at	Ground.	Frequency.	Observations.
	fathoms.	fathoms.			
Lysianassa, *M.-Edw.*					
Audouiniana, *Spence Bate*					
longicornis, *Lucas* (Chausica,					
Spence Bate, not *M.-Edwards*)					
Atlantica, *Edw.* (marina, *Spence*					
Bate)					
Callisoma, *Hope* (Scopelocheirus,					
Spence Bate).					
crenata, *Spence Bate*					
Anonyx, *Kröyer.*					
Edwardsii, *Kröyer*					
minutus, *Kröyer*					
Holbölli, *Kröyer* ....!					
ampulla, *Kröyer*					
denticulatus, *Spence Bate*......					
longipes, *Spence Bate*					
obesus, *Spence Bate*.........					
longicornis, *Spence Bate*					
Opis, *Kröyer.*					
typica, *Kröyer*.............					
Ampelisca, *Kröyer* (Tetromatus,					
Spence Bate).					
Gaimardii, *Kröy.* (typica, *Sp.B.*)					
Belliana, *Spence Bate*					
Westwoodilla (Westwoodia,					
Spence Bate).					
cæcula, *Spence Bate*					
hyalina, *Spence Bate*					
Monoculodes. *Stimpson.*					
carinatus, *Spence Bate*					
Kröyera, *Spence Bate.*					
arenaria, *Spence Bate*					
Phoxus, *Kröyer.*					
simplex, *Spence Bate* (Kröyeri,					
Spence Bate, not *Stimpson*)..					
plumosus, *Holböll* ....!					
Holbölli, *Kröy.*					
Sulcator, *Spence Bate.*					
arenarius, *Spence Bate*					
Urothoë, *Dana.*					
marinus, *Spence Bate* (Sulcator					
marinus) ....!					
Bairdii, *Spence Bate*					
medius, *Spence Bate*					
elegans, *Spence Bate*					
Gratia, *Spence Bate.*					
imbricata, *Spence Bate*					
Liljeborgia, *Spence Bate.*					
pallida, *Spence Bate*					

25

CRUSTACEA.

Species.	Range.	Found living at	Ground.	Frequency.	Observations.
	fathoms.	fathoms.			
Phliora, *Spence Bate.*					•
antiqua, *Spence Bate*					
Kinahani, *Spence Bate.*					
Isæa, *M.-Edwards.*					
Montagui, *M.-Edw.*					
Iphimedia, *Rathke.*					
obesa, *Rathke*					
Eblanæ. *Spence Bate*					
Otus. *Spence Bate.*					
carinatus, *Spence Bate* .	.				
Acanthonotus, *Owen.*					
testudo, *Montagu*					
Dexamine, *Leach.*					
Loughrinii. *Spence Bate*					
spinosa, *Mont.*					
Ersirus. *Kröyer.*					
Edwardi. *Spence Bate*	′.				
Atylus, *Leach.*					
bispinosus, *Spence Bate*					
Huxleyanus, *Spence Bate*					
Gordonianus, *Spence Bate*					
Phærusa, *Leach.*					
cirrus, *Spence Bate* . . .	.				
fucicola, *Edw.*					
Calliope, *Leach.*					
Leachii, *Spence Bate*					
Ersirus, *Spence Bate.*					
Helvetiæ, *Spence Bate* . .					
Lemnos, *Spence Bate.*					
Cambriensis, *Spence Bate*					
versiculatus. *Spence Bate*					
Websterii, *Spence Bate* . . .					
Danmoniensis, *Spence Bate*					
Aora, *Kröy.* (=Lalaria, *Nicolet*).					
gracilis, *Spence Bate*					
Ecrystheus, *Spence Bate.*					
tridentatus, *Spence Bate* .					
tuberculosus. *Spence Bate* .					
Gammarella, *Spence Bate.*					
brevicaudata. *M.-Edw.*(=G. orchestiformis, *Spence Bate*.)					
Crangonyx, *Spence Bate.*					
subterranea, *Spence Bate* .					
Amathia. *Rathke.*					
Sabinii. *Leach*					
Gammarus. *Fabr.*					
locusta. *Fabr.*				•	
fluviatilis, *Rosel*					
gracilis. *Rathke*					

26

Species.	Range.	Found living at	Ground.	Frequency.	Observations.
	fathoms.	fathoms.			
GAMMARUS, *Fabr.*					
camptolops, *Leach*					
marinus, *Leach*					
laminatus, *Johnston*					
longimanus, *Leach*					
palmatus, *Mont.* (inæquimanus, *Spence Bate*)					
grossimanus, *Mont.*					
maculatus, *Johnston*					
BATHYPOREIA, *Lindström* (THERSITES, *Spence Bate*).					
pilosa, *Lindström*					
pelagica, *Spence Bate*					
Robertsoni, *Spence Bate*					
LEUCOTHOË, *Leach*, not *Kröyer.*					
articulosa, *Mont.*					
furina, *Savig.* (procera, *Sp. Bate*)					
PLEONEXES, *Spence Bate.*					
gammaroides, *Spence Bate*					
AMPHITHOË, *Leach.*					
rubricata, *Mont.*					
littorina, *Spence Bate*					
? obtusata, *Leach*					
? dubia, *Johnston*					
SUNAMPHITHOË, *Spence Bate.*					
hamulus, *Spence Bate*					
conformata, *Spence Bate*					
PODOCERUS, *Leach.*					
falcatus, *Mont.*	●				
variegatus, *Leach*					
pulchellus, *Leach*					
JASSA ?, *Leach.*					
pelagica, *Leach*					
SIPHONŒCETUS, *Kröyer.*					
Whitei, *Gosse*					
ERICHTHONIUS, *M.-Edw.*	●				
difformis, *M.-Edw.*			●		
CYRTOPHIUM, *Dana.*					
Darwinii, *Spence Bate*					
COROPHIUM, *Latreille.*					
longicorne, *Fabr.*					
CHELURA, *Philippi.*					
terebrans, *Phil.*					
HYPERIA, *Latreille.*					
Galba, *Mont.* (Latreillü, *Edw.* = Metoëcus medusarum, *Latr.*)					
oblivia, *Kröy.*					

27

CRUSTACEA.

AMPHIPODA HYPERINA.

Species.	Range.	Found living at	Ground.	Fre-quency.	Observations
	fathoms.	fathoms.			
Læstrigonus, *Guérin*.					
Fabricii, *M.-Edw.*					
Phronima, *Latr.*					
sedentaria, *Forsk.*					
Typhis, *Risso.*					
nolens, *Johnston*					

AMPHIPODA ABERRANTIA.   (Læmodipoda of *Latreille.*)

Dulichia, *Kröyer.*
  porrecta, *Spence Bate*
  falcata, *Spence Bate*
Proto, *Leach.*
  pedata, *Leach* ....
  Goodsirii, *Spence Bate* ....
Protella, *Dana.*
  longispina, *Kröyer* ..
Caprella, *Lamarck.*
  linearis, *Latr.* ....
  Pennantii, *Leach* ....
  tuberculosa, *Goodsir* ....
  lobata, *Müller* ....
  acuminifera, *M.-Edw.* ....
Cyamus, *Latreille.*
  ceti, *Linn.* ....
  ovalis, *Roussel* ....
  gracilis, *Roussel* ....
  Thompsoni, *Gosse*

ISOPODA ABERRANTIA.   (Anisopoda of *Dana.*)

Arcturus, *Latr.* (Astacilla, *John-*
    ston; Leachia, *Johnston.*)
  longicornis, *Sow.* ....
  intermedius, *Goodsir* ....
  gracilis, *Goodsir* ....
Anthura, *Leach.*
  gracilis, *Mont.* ....
  cylindricus, *Mont.* ......
Tanais, *M.-Edw.*
  Dulongii, *And.* ... .
  hirticaudatus, *Spence Bate*
28

CRUSTACEA.

Species.	Range	Found living at	Ground.	Frequency.	Observations.
	fathoms.	fathoms.			
Apseudes, *Leach.*					
talpa, *Mont.*					
Anceus, *Risso.*					
maxillaris, *Mont.*					
rapax, *M.-Edw.*					
Praniza, *Leach.*					
ceruleata, *Mont.*					
fusca ?, *Johnston*					
Edwardii, *Spence Bate*					
Liriope, *Kröyer.*					
balani, *Spence Bate*					
Ione, *Mont.*					
thoracica, *Mont.*					
Bopyrus, *Latr.*					
squillarum, *Latr.*					
hippolytes, *Kröyer*					
Phryxus, *Rathke.*					
hippolytes, *Rathke*					
paguri, *Rathke.*					

ISOPODA (NORMALIA).

Munna, *Kröyer.*					
Kröyeri, *Goodsir*					
Whiteana, *Spence Bate*					
Jæra, *Leach.*					
albifrons, *Leach*					
Oniscoda, *Latreille.*					
maculosa, *Leach*					
Deshayesii, *Lucas*					
Limnoria, *Leach.*					
lignorum, *Rathke* (terebrans, *Leach*)			●		
Idotea, *Fabr.*					
pelagica, *Leach*					
tricuspidata, *Desm.*					
emarginata, *Fabr.*					
linearis, *Latr.*					
acuminata, *Leach*					
appendiculata, *Risso*					
Ligia, *Fabr.*					
oceanica, *Linn.*					
Sphæroma, *Latr.*					
serratum, *Fabr.*					
rugicauda, *Leach*					
Hookeri, *Leach*					

29

c

CRUSTACEA.

Species.	Range	Found living at	Ground.	Frequency.	Observations.
	fathoms.	fathoms.			
CYMODOCEA, *Leach.*					
truncata, *Leach*					
emarginata, *Leach*					
Montagni, *Leach*					
rubra, *Leach*					
viridis, *Leach*					
NERLEA, *Leach.*					
bidentata, *Adams*					
CAMPECOPEA, *Leach.*					
hirsuta, *Mont.*					
Cranchii, *Leach*					
CIROLANA, *Leach.*					
Cranchii, *Leach*					
EURYDICE, *Leach.*					
pulchra, *Leach*					
ÆGA, *Leach.*					
bicarinata, *Leach*					
tridens, *Leach*					
CONILERA, *Leach.*					
cylindracea, *Mont.*					
ROCINELA, *Leach.*					
Dannioniensis, *Leach*					
monophthalma, *Johnston*					

ENTOMOSTRACA.

Order I. PHYLLOPODA.

NEBALIA, *Leach.*
bipes, *O. Fabr.*
ARTEMIA, *Leach.*
salina, *Linn.*

Order II. CLADOCERA.

EVADNE, *Lovén.*
Nordmanni, *Lovén*

Order III. OSTRACODA.

Fam. I. **Cytheridæ.**

CYTHERE, *Müller.*
flavida, *Müll.*
reniformis, *Baird*
albo-maculata, *Baird*
alba, *Baird*

30

Species.	Range.	Found living at	Ground.	Fre-quency.	Observations
	fathoms.	fathoms.			
CYTHERE, *Müller.*					
variabilis, *Baird*					
aurantia, *Baird*					
nigrescens, *Baird*					
Minna, *Baird*					
angustata, *Münster* ..					
acuta, *Baird*					
pellucida, *Baird*					
impressa, *Baird*					
quadridentata, *Baird*					
convexa, *Baird*					
CYTHEREIS, *Rupert Jones.*					
Whitei, *Baird*					
Jonesii, *Baird*					
antiquata, *Baird*					
Fam. II. **Cypridinidæ.**					
CYPRIDINA, *M.-Edw.*					
Macandrei, *Baird*					
Brenda, *Baird*					
Mariæ, *Baird*					
interpuncta, *Baird*					
Order IV. COPEPODA.					
Fam. 1. **Cyclopidæ.**					
CANTHOCAMPTUS, *Westwood.*					
Strömii, *Baird* ..:					
furcatus, *Baird*					
minuticornis, *Müll.* ...					
ARPACTICUS, *M.-Edw.*					
chelifer, *Müll.*					
nobilis, *Baird*					
ALTEUTHA, *Baird.*					
depressa, *Baird*					
Fam. II. **Diaptomidæ.**					
TEMORA, *Baird.*					
Finmarchica, *Gunner*					
CALANUS.					
euchæta, *Lubbock*					
Anglicus, *Lubbock* ..					
ANOMALOCERA, *Templeton.*					
Patersonii, *Temple.* ..					

31

Species.	Range.	Found living at	Ground.	Frequency.	Observations.
	fathoms.	fathoms.			
Fam. III. Cetochilidæ.					
CETOCHILUS, *Vauzème.*					
septentrionalis, *Goodsir*	..				
PONTELLA, *Dana.*					
Wollastoni, *Lubbock*	....				
PONTELLINA, *Dana.*					
brevicornis, *Lubbock*					
PELTIDIUM, *Philippi.*					
purpureum °, *Phil.*	....				
CORYCÆUS, *Dana.*					
Anglicus, *Lubbock*	....				
Fam. IV. Monstrillidæ.					
MONSTRILLA, *Dana.*					
Anglica, *Lubbock*					

Species.	On what Animals found.	Observations.
Order V. SIPHONOSTOMA.		
Fam. I. Caligidæ.		
CALIGUS, *Müller.*		
diaphanus, *Nordm.*	Various fishes.	
rapax, *M.-Edw.*	Various fishes.	
Mülleri, *Leach*	Various fishes.	
centrodonti, *Baird*	Sea Bream.	
minutus, *Otto*	Holibut.	
curtus, *Müll.*	Ray.	
LEPEOPHTHEIRUS, *Nordmann.*		
Strömii, *Baird*	On Salmon.	
pectoralis, *Müll.*	Various fishes.	
Nordmanni, *M.-Edw.*	Sun-fish.	
hippoglossi, *Kröy.*	Holibut.	
obscurus, *Baird*	Brill.	
Thompsoni, *Baird*	Turbot.	
CHALIMUS, *Burmeister.*		
scombri, *Burm.*	On Mackerel.	
TREBIUS, *Kröyer.*		
caudatus, *Kröy.*	On Skate.	
Fam. II. Pandaridæ.		
DINEMOURA, *Latreille.*		
alata, *M.-Edw.*	On Shark.	
lamnæ, *Johnst*	On Shark.	

32

Species.	On what Animals found.	Observations.
Fam. III. Cecropidæ.		
PANDARUS, *Leach*.		
bicolor, *Leach*	On Shark.	
CECROPS, *Leach*.		
Latreillei, *Leach*	On Sun-fish.	
LEMARGUS, *Kröyer*.		
muricatus, *Kröy.*	On Sun-fish.	
Fam. IV. Anthosomidæ.		
ANTHOSOMA, *Leach*.		
Smithii, *Leach*	On Shark.	
Fam. V. Ergasilidæ.		
NICOTHOË, *M.-Edwards*.		
astaci, *M.-Edw.*	On gills of Lobster	
Fam. VI. Chondracanthidæ.		
CHONDRACANTHUS, *De la Roche*.		
Zei. *De la Roche*	Gills of Dory.	
LERNENTOMA, *De Blainville*.		
cornuta, *Müll.*	Gills of Sole.	
asellina, *L.*	Gills of Gurnard.	
lophii, *Johnst.*	Pouches of Angler.	
LERNEOPODA, *De Blainville*.		
elongata, *Grant*	Shark.	
Galei, *Kröy.*	Shark.	
salmonea, *L.*	Salmon.	
Fam. VII. Anchorellidæ.		
ANCHORELLA, *Cuvier*.		
uncinata, *Müll.*	On the Cod.	
rugosa, *Kröy.*	On the Cod.	
Fam. VIII. Lerneidæ.		
LERNEA, *L.*		
branchialis, *L.*	Gills of Cod.	
LERNEONEMA, *M.-Edwards*.		
sprattæ, *Sby.*	On the Sprat.	
Bairdii, *Salter*	Herring.	
encrasicoli, *Turt.*	Sprat.	

33

CIRRIPEDIA.

Species.	Range.	Found living at	Ground.	Fre-quency.	Observations.
	fathoms.	fathoms.			
Fam. I. Balanidæ.					
BALANUS (*Lister*).					
tintinnabulum, *Linn.*					
spongicola, *Brown*					
perforatus, *Bruguière*	.				
Amphitrite, *Darwin*					
eburneus, *Aug. Gould*					
improvisus, *Darwin*					
porcatus, *Da Costa*					
crenatus, *Bruguière*					
balanoides, *Linn.*					
Hameri, *Ascanius*					
ACASTA, *Leach.*					
spongites, *Poli*					
PYRGOMA, *Leach.*					
Anglicum, *G. B. Sby.*					
XENOBALANUS, *Steenstrup.*					
globicipitis, *Steenstrup*					
CHTHAMALUS, *Ranzani.*					
stellatus, *Poli*					
VERRUCA, *Schumacher.*					
Strömia, *O. Müller*					
ALCIPPE, *Hancock.*					
lampas, *Hancock*					
Fam. II. Lepadidæ.					
LEPAS, *Linn.*					
anatifera, *Linn.*					
Hillii, *Leach*					
anserifera, *Linn.*					
pectinata, *Spengler*					
fascicularis, *Ellis & Solander* ..					
CONCHODERMA, *Olfers.*					
aurita, *Linn.*					
virgata, *Spengler*					
ALEPAS, *Sander-Rang.*					
parasita, *Sander-Rang*					
ANELASMA, *Darwin.*					
squalicola, *Lovén*					
SCALPELLUM, *Leach.*					
vulgare, *Leach*	.				
POLLICIPES, *Leach.*					
cornucopia, *Leach*					

34

ARACHNIDA.

Species.	Range.	Found living at	Ground.	Fre-quency.	Observations.
	fathoms.	fathoms.			
Order PODOSOMATA.					
Fam. I. Pycnogonidæ.					
PYCNOGONUM, *Fabricius.*					
littorale, *Strom.*					
PHOXICHILUS, *Latreille.*					
spinosus, *Mont.*.............					
Fam. II. Nymphonidæ.					
PHOXICHILIDIUM, *M.-Edwards.*					
coccineum, *Johnst.*					
globosum					
olivaceum					
PALLENE, *Johnston.*					
brevirostris, *Johnst.*					
NYMPHON, *Fabricius.*					
gracile, *Leach*					
grossipes, *O. Fabr*...........					
femoratum, *Leach*...........					
pictum					
giganteum, *Johnst.*					

ANELLIDA.

. The following list of British Marine Worms is copied, by favour of Dr. J. E. Gray, from an unpublished Catalogue by the late Dr. Johnston.

Order I. TURBELLARIA.					
Fam. I. Planoceridæ.					
LEPTOPLANA, *Ehrenberg.*					
subauriculata, *Johnston*					
tremellaris, *Müller*					
flexilis, *Dalyell*					
atomata, *Müller*					
ellipsis, *Dalyell*					
EURYLEPTA, *Ehrenberg.*					
cornuta, *Müller*					
Dalyellii, *Johnston*					
sanguinolenta, *Quatrefages*					
vittata, *Montagu*					
PLANOCERA, *Blainville.*					
folium, *Grube*					

35

Species.	Range.	Found living at	Ground.	Frequency.	Observations.
	fathoms	fathoms.			
Fam. 11. **Planariadæ.**					
PLANARIA, *Müller.*					
ulvæ, *Oersted*					
affinis, *Oersted*					
alba, *Dalyell*					
variegata, *Dalyell*					
? gracilis, *Dalyell*					
? falcata, *Dalyell*					
Fam. III. **Dalyellidæ.**					
TYPHLOPLANA, *Ehrenberg.*					
flustræ, *Dalyell*					
CONVOLUTA, *Oersted.*					
paradoxa, *Oersted*					
Doubtful species of this family.					
PLANOIDES fusca, *Dalyell*					
PLANARIA hirudo, *Johnston*					
ASTEMMA, *Oersted.*					
rufifrons, *Johnston*					
filiformis, *Johnston*					
CEPHALOTRIX, *Oersted.*					
lineatus, *Dalyell*					
flustræ, *Dalyell*					
TETRASTEMMA, *Ehrenberg.*					
varicolor, *Oersted*					
variegatum, *Dalyell*					
? algæ, *Dalyell*					
BORLASIA, *Johnston.*					
olivacea, *Johnston*					
octoculata, *Johnston*					
purpurea, *Johnston*					
Gesserensis, *Müller*					
striata, *Rathke*					
OMMATOPLEA, *Ehrenberg.*					
gracilis, *Johnston*					
rosea, *Müller*					
alba, *Thompson*					
melanocephala, *Johnston*					
pulchra, *Johnston*					
SYLLUS, *Johnston.*					
viridis, *Dalyell*					
purpureus, *Dalyell*					
fragilis, *Dalyell*					
fasciatus, *Dalyell*					
LINEUS, *T. W. Simmons.*					
longissimus, *Simmons*					

36

ANELLIDA.

Species.	Range.	Found living at	Ground.	Fre-quency.	Observations
	fathoms.	fathoms.			
LINEUS, *T. W. Simmons.*					
gracilis, *Goodsir*					
lineatus, *Johnston*					
murenoides, *D. Chiaje*					
fasciatus, *Johnston*					
viridis, *Dalyell*.............					
albus, *Dalyell*					
MECKELIA, *Leuckart.*					
annulata, *Montagu*					
tænia, *Dalyell*					
SERPENTARIA, *H. D. S. Goodsir.*					
fragilis, *Goodsir*					
fusca, *Dalyell*					
Order II. BDELLOMORPHA.					
Fam. 1. **Malacobdellidæ.**					
MALACOBDELLA, *Blainville.*					
grossa, *Müller*					
Valenciennæi, *Blanchard*......					
anceps, *Dalyell*					
Order III. BDELLIDEA.					
Fam. I. **Branchelliadæ.**					
BRANCHELLION, *Savigny.*					
torpedinis, *Savigny*					
Fam. II. **Piscicolidæ.**					
PONTOBDELLA, *Leach.*					
muricata, *Linn.*					
verrucata, *Grube*					
arcolata, *Leach*					
lævis, *Blainville*					
littoralis, *Johnston*					
campanulata, *Dalyell*					
Order IV. SCOLICES.					
Fam. I. **Lumbricidæ.**					
SÆNURIS, *Hoffmeister.*					
lineata, *Müller*.............					
CLITELLIO, *Savigny.*					
arenurius, *Müller*...........					
VALLA, *Johnston.*					
ciliata, *Müller*					

37

ANELLIDA.

Species.	Range.	Found living at	Ground.	Frequency.	Observations.
	fathoms.	fathoms.			
Order V. GYMNOCOPA.					
Fam. I. Tomopteridæ.					
TOMOPTERIS, *Eschscholtz.*					
onisciformis, *Grube*					
Order VI. CHÆTOPODA.					
Fam. I. Aphroditidæ.					
APHRODITA, *Leach.*					
aculeata, *Linn.*					
borealis, *Johnston*					
hystrix, *Savigny*					
LEPIDONOTUS, *Leach.*					
squamatus, *Linn.*					
clava, *Montagu*					
impar, *Johnston*					
pharebratus, *Johnston*					
cirratus, *Fabr.*					
semisculptus, *Leach*					
pellucidus, *Dyster*					
imbricatus, *Linn.*					
Species not defined.					
APHRODITA squamata, *Dalyell* ..					
lepidota, *Pallas*					
minuta, *Pennant*					
annulata, *Pennant*					
velox, *Dalyell*					
LEPIDONOTUS floccosus ?, *Dalyell.*					
POLYNOE semisquamosa, *Williams*					
POLYNOË, *Oersted.*					
scolopendrina, *Savigny*					
PHOLOË, *Johnston.*					
inornata, *Johnston*					
eximia, *Dyster*					
SIGALION, *Aud. & M.-Edwards.*					
boa, *Johnston*					
Fam. II. Amphinomenidæ.					
EUPHROSYNE, *Savigny.*					
foliosa, *Aud. & M.-Edwards*....					
borealis, *Oersted*					
Fam. III. Euniceidæ.					
EUNICE, *Aud. & M.-Edwards.*					
Norvegica, *Linn.*					

38

ANELLIDA.

Species.	Range.	Found living at	Ground.	Fre-quency.	Observations.
	fathoms.	fathoms.			
EUNICE, *Aud. & M.-Edwards.*					
annulicornis, *Brit. Mus.*					
antennata, *Savigny*					
Harassii, *Aud. & M.-Edwards*..					
sanguinea, *Montagu*					
margaritacea, *Williams*					
NORTHIA, *Johnston.*					
tubicola, *Müller*					
conchylega, *Sars*					
LYCIDICE, *Savigny.*					
Ninetta, *Aud. & M.-Edwards* ..					
rufa, *Gosse*					
LUMBRINERIS, *Blainville.*					
tricolor. *Leach*					
Fam. IV. **Nereidæ.**					
NEREIS, *Cuvier.*					
brevimana, *Johnston*.........					
pelagica, *Linn.*.............					
diversicolor, *Müller*					
cœrulea, *Linn.*.............					
fimbriata, *Müller*					
imbecillis, *Grube*					
Dumerilii, *Aud. & M.-Edwards*					
pulsatoria, *Montagu*.........					
NEREILEPAS, *Oersted.*					
fucata, *Savigny*					
HETERONEREIS, *Oersted.*					
lobulata, *Savigny*...........					
renalis, *Johnston*					
longissima, *Johnston*.........					
margaritacea, *Johnston*.......					
Fam. V. **Nephthyidæ.**					
NEPHTHYS, *Cuvier.*					
cœca, *Fabr.*					
longisetosa, *Oersted*					
Hombergii, *Cuv.*??					
Fam. VI. **Phyllodoceidæ.**					
PHYLLODOCE, *Cuvier.*					
lamelligera, *Turton*					
bilineata, *Johnston*					
maculata, *Linn.*					
viridis, *Linn.*					
ellipsis, *Dalyell*					

39

Species.	Range.	Found living at	Ground.	Frequency	Observations.
	fathoms.	fathoms.			
PHYLLODOCE, *Cuvier.*					
Griffithsii, *Johnston*					
cordifolia, *Dyster*					
PSAMATHE, *Johnston.*					
punctata, *Müller*					
Fam. VII. Glyceridæ.					
GLYCERA, *Savigny.*					
mitis, *Johnston*					
dubia, *Blainville*					
capitata, *Oersted*					
nigripes, *Johnston*					
GONIADA, *Aud. & M.-Edwards.*					
maculata, *Oersted*					
Fam. VIII. Syllidæ.					
SYLLIS, *Savigny.*					
armillaris, *Müller*					
cornuta, *H. Rathke*					
prolifera, *Müller*					
? monoceros, *Dalyell*					
GATTIOLA, *Johnston.*					
spectabilis, *Johnston*					
MYRIANIDA, *M.-Edw.*					
pinnigera, *Montagu*					
IOIDA, *Johnston.*					
macrophthalma, *Johnston*					
Fam. IX. Amytideidæ.					
AMYTIDEA, *Grube.*					
maculosa (NEREIS), *Montagu* ..					
Fam. X. Ariciadæ.					
NERINE, *Johnston.*					
vulgaris, *Johnston*					
coniocephala, *Johnston*					
Doubtful species.					
NERINE contorta (NEREIS), *Daly.* }					
SPIO, *Turton.*					
filicornis, *Müller*·					
seticornis, *Turton*					
crenaticornis, *Montagu*					
LEUCODORE, *Johnston.*					
ciliatus, *Johnston*					

40

ANELLIDA.

Species.	Range.	Found living at	Ground.	Fre-quency.	Observations.
	fathoms.	fathoms.			

EPHESIA, *Rathke*.
  gracilis, *Rathke* ....
SPHÆNODONUM, *Oersted*.
  peripatus, *Johnst*...........
CIRRATULUS, *Lamarck*.
  tentaculatus, *Mont*..........
  borcalis, *Lamk*. ....
DODECACERIA, *Oersted*.
  concharum, *Oersted* ....

Fam. XI. **Opheliadæ.**

OPHELIA, *Savigny*.
  acuminata, *Oersted* ....
AMMOTRYPANE, *Rathke*.
  limacina, *Rathke* ....
TRAVISIA, *Johnston*.
  Forbesii, *Johnst*. ....
EUMENIA, *Oersted*.
  crassa, *Oersted* ....

Fam. XII. **Siphonostomidæ.**

SIPHONOSTOMA, *Cuvier*.
  uncinata, *Aud. & M.-Edw*.....
TROPHONIA, *Cuvier*.
  plumosa, *Müller* ... .

Fam. XIII. **Telethusidæ.**

ARENICOLA, *Savigny*.
  piscatorum, *Lamk*. ....
  branchialis, *Aud. & M.-Edw*...
  ccaudata, *Johnst*............

Fam. XIV. **Maldaniadæ.**

CLYMENE, *Savigny*.
  borealis, *Dalyell* ....

Fam. XV. **Terebellidæ.**

TEREBELLA, *Montagu*.
  conchilega, *Pallas* ....
  littoralis, *Dalyell* ....
  cirrata, *Mont*. ....
  nebulosa, *Mont*..............
  gigantea, *Mont*..............
  constrictor, *Mont*............
  venustula, *Mont*. ....
  tuberculata, *Dalyell* ....
  textrix, *Dalyell* ....
  maculata, *Dalyell*...........

41

Species.	Range.	Found living at	Ground.	Frequency.	Observations.
	fathoms.	fathoms.			
VENUSIA, *Johnston.*					
punctata, *Johnst.* .					
TEREBELLIDES, *Sars.*					
Strœmii, *Sars*					
PECTINARIA, *Lamarck.*					
Belgica, *Pallas*					
granulata, *Linn.*					
Fam. XVI. **Sabellariadæ.**					
SABELLARIA, *Lamarck.*					
Anglica, *Ellis*					
crassissima, *Leach.*					
lumbricalis, *Mont.*					
Fam. XVII. **Serpulidæ.**					
ARIPPASA, *Johnston.*					
infundibulum, *Mont.*					
SABELLA, *Savigny.*					
pavonina, *Savigny*					
penicillus, *Linn.*					
vesiculosa, *Mont.*					
bombyx, *Dalyell*					
Savignii, *Johnst.*					
volutacornis, *Mont.*					
Doubtful Species.					
SABELLA unispira, *Sav.* .					
rosea (AMPHITRITE), *Sow.*					
luna (AMPHITRITE), *Dalyell* .					
curta (AMPHITRITE), *Mont.* .					
PROTULA, *Risso.*					
protensa, *Philippi*					
Dysteri, *Huxley*					
SERPULA, *Linnæus.*			.		
vermicularis, *Ellis* ..					
intricata, *Linn.*					
reversa, *Mont.*					
Berkeleyi, *Johnst.*					
conica, *Flem.*					
armata, *Flem.*					
Dysteri, *Johnst.*					
DITRUPA, *Berkeley.*					
subulata, *Deshayes*					
FILOGRANA, *Berkeley.*					
implexa, *Beck.*					
OTHONIA, *Johnston.*					
Fabricii, *Johnst.*					

42

Species.	Range.	Found living at	Ground.	Fre-quency.	Observations.
	fathoms.	fathoms.			
Fam. XVIII. **Campontiadæ**.					
CAMPONTIA, *Johnston*.					
cruciformis, *Johnst*.					
Fam. XIX. ? **Mæadæ**.					
MÆA. *Johnston*.					
mirabilis, *Johnst*.					
Fam. XX. ? **Sipunculidæ**.					
SYRINX, *Bohadsch*.					
nudus, *Linn*.					
papillosus, *Thomps*.					
Harveii, *Forbes*					
SIPUNCULUS, *Linnæus*.					
Bernhardus, *Forbes*					
MACRORHYNCHOPTERUS, *Rondel*.					
Johnstoni, *Forbes*					
saccatus, *Flem*.					
tenuicinctus, *M'Coy*					
Forbesii, *M'Coy*					
granulosus, *M'Coy*					
Pallasii, *Forbes*					
Fam. XXI. **Priapulidæ**.					
PRIAPULUS, *Lamarck*.					
caudatus, *Flem*.					
Fam. XXII. **Thalassemidæ**.					
THALASSEMA, *Cuvier*.					
Neptuni, *Gaërtner*					
ECHIURUS, *Cuvier*.					
oxyurus, *Pall*.					
Species inquirendæ.					
NEREIS, *Cuvier*.					
iricolor, *Mont*.					
margarita, *Mont*.					
lineata, *Mont*.					
maculosa, *Mont*.					
rufa, *Penn*.					
mollis, *Linn*.					
octentaculata, *Mont*.					
punctata, *Encycl. Méth*.					
noctiluca, *Linn*.					
pinnigera, *Mont*.					
APHRODITA, *Leach*.					
annulata, *Penn*.					
minuta, *Penn*.					

43

F 2

Species.	Range.	Found living at	Ground.	Fre-quency.	Observations.
	fathoms.	fathoms.			

Species inquirendæ.

Spio, *Turton.*
  seticornis, *Turt.* . . . .
  crenaticornis, *Mont.* . . . .
  calcarea, *Templeton* . . . .
Branchiarius, *Mont.*
  quadrangularis. *Mont.* . . . .
Diplotis, *Mont.*
  hyalina, *Mont.* . . . .
Derris, *Adams.*
  sanguinea. *Adams* . . . .

ENTOZOA.

₊ From Dr. Baird's British Museum Catalogue; and Dr. Bellingham's List of Irish Entozoa, in the 'Annals of Natural History,' 1844.

Species.	In what Animals found.	Observations.
Order NEMATOIDEA.		
Fam. **Filariadæ.**		
Filaria, *Müller.*		
? marina, *Linn.*	Shad and Cod.	
inflexo-caudata, *Siebold*	Porpoise.	
sp.	Red Gurnard and Mullet.	
Trichosoma, *Rudolphi.*		
gracilis, *Bellingh.*	Hake.	
Spiroptera, *Rudolphi.*		
sp.	Skate.	
Fam. **Ascaridæ.**		
Ascaris, *Linnæus.*		
osculata, *Rud.*	Seal.	
ancta, *Rud.*	Viviparous Blenny.	
rigida, *Rud.*	Lophius.	
capsularia, *Rud.*	Cod. &c.	
collaris, *Rud.*	Flounder. &c.	
clavata, *Rud.*	Cod. &c.	
constricta, *Rud.*	Sea-Scorpion. &c.	
rotundata, *Rud.*	Skate.	

44

Species.	In what Animals found.	Observations.
Ascaris, *Linnœus.*		
acus, *Bloch*	Herring.	
angulata, *Rud.*	Lophius.	
tenuissima, *Zeder*	Whiting.	
succisa, *Rud.*	Lump-fish.	
Fam. Sclerostomidæ.		
Cucullanus, *Müller.*		
minutus, *Rud.*	Flounder.	
heterochrous, *Rud.*	Flounder.	
foveolatus, *Lam.*	Plaice and Dab.	
Stenurus, *Dujardin.*		
inflexus, *Rud.* (part.)	Porpoise.	
Prosthecosacter, *Diesing.*		
inflexus, *Rud.*	Porpoise.	
convolutus, *Kuhn*	Porpoise.	
Order TREMATODA.		
Fam. Onchobothriadæ.		
Octobothrium, *Leuckart.*		
lanceolatum, *Leuck.*	Shad.	
Fam. Capsalidæ.		
Capsala, *Bosc.*		
coccinea. *Cuv.*	Sun-fish.	
elongata. *Nitzsch*	Sturgeon.	
Fam. Distomidæ.		
Monostoma, *Zeder.*		
filicolle, *Rud.*	Sea Bream.	
trigonocephalum, *Rud.*	Turtle.	
Distoma, *Retzius.*		
appendiculatum, *Rud.*	Shad. &c.	
hispidum, *Viborg*	Sturgeon.	
mogastomum, *Rud.*	Smooth Shark.	
microcephalum, *Baird*	Spinous Shark.	
tumidulum, *Rud.*	Pipe-fish.	
fulvum, *Rud.*	Skate.	
varicum	Salmon.	
gibbosum ?, *Rud.*	Haddock.	
rufo-viride. *Rud.*	Conger-eel.	
reflexum ?	Cyclopterus.	
excisum, *Rud.*	Mackerel.	
scabrum. *Zeder*	Whiting.	
contortum, *Rud.*	Sun-fish.	
nigro-flavum, *Rud.*	Sun-fish.	

45

ENTOZOA.

Species.	In what Animals found.	Observations.
HIRUDINELLA, *Garsin.*		
clavata, *Menzies*	Bonito.	
Order ACANTHOCEPHALA.		
Fam. **Echinorhynchidæ.**		
ECHINORHYNCHUS, *Müller.*		
proteus, *Westrumb*	Flounder.	
aeus, *Rud.*	Cod, &c.	
gibbosus, *Rud.*	Herring.	
strumosus, *Rud.*	Seal.	
Order CESTOIDEA.		
Fam. **Rhynchobothridæ.**		
RHYNCHOBOTHRIUM, *Blainville.*		
corollatum, *Abildg.*	Smooth Shark.	
TETRARHYNCHUS, *Rudolphi.*		
megacephalus, *Rud.*	Spotted Dog-fish.	
solidus, *Drum.*	Salmon.	
grossus, *Rud.*	Salmon.	
rugosus, *Baird*	Salmon.	
TETRABOTHRIORHYNCHUS, *Diesing.*		
barbatus, *Linn.*	Lemon Sole.	
Fam. **Tæniadæ.**		
BOTHRIOCEPHALUS, *Rudolphi.*		
fragilis, *Rud.*	Shad.	
proboscideus, *Batsch*	Salmon, &c.	
punctatus, *Rud.*	Turbot, &c.	
tumidulus, *Rud.*	Ray.	
microcephalus, *Rud.*	Sun-fish.	
coronatus, *Rud.*	Skate.	
corollatus, *Rud.*	Dog-fish.	
paleaceus, *Rud.*	Dog-fish.	
Fam. **Scolecidæ.**		
SCOLEX, *Müller.*		
polymorphus, *Rud.*	Turbot, &c.	
Order CYSTICA.		
Fam. **Cysticidæ.**		
ANTHOCEPHALUS, *Rudolphi.*		
elongatus, *Rud.*	Sun-fish.	
granulosus?, *Rud.*	Whiting, &c.	
paradoxus, *Drum.*	Turbot.	

46

CLASS ECHINODERMATA.

Species.	Range.	Found living at	Ground.	Frequency.	Observations.
	fathoms.	fathoms.			

Order CRINOIDEA.

COMATULA, *Lamarck*.
  rosacea, *Link* . . . .
  Celtica, *Barrett* . . . .
  Sarsii, *Von Düben & Koren* . . . .

Order OPHIUROIDEA.

Fam. **Ophiuridæ.**

OPHIURA, *Lamarck*.
  texturata, *Lamk.* . . . .
  albida, *Forbes* . . . .
OPHIOCOMA. *Agassiz.*
  neglecta, *Johnst.* . . . .
  punctata, *Forbes* . . . .
  filiformis, *Müller* . . . .
  securigera, *D. & K.* . . . .
  bellis, *Link* . . . .
  brachiata, *Mont.* . . . .
  Ballii, *Thomps.* . . . .
  Goodsiri, *Forbes* . . . .
  granulata, *Link* . . . .
  rosula, *Link* . . . .
  minuta, *Forbes.* . . . .

Subfam. **Euryalidæ.**

ASTROPHYTON, *Link.*
  scutatum, *Link* . . . .

Order ASTEROIDEA.

Fam. **Asteriadæ.**

URASTER, *Agassiz.*
  glacialis, *Linn.* . . . .
  rubens, *Linn.* . . . .
  violacea, *Muller* . . . .
  hispida, *Penn.* . . . .
  rosea, *Müller* . . . .
ECHINASTER, *Müller & Troschel.*
  oculatus, *Penn.* . . . .
SOLASTER, *Forbes.*
  endeca, *Linn.* . . . .
  papposa, *Linn.* . . . .

47

ECHINODERMATA.

Species.	Range.	Found living at	Ground.	Frequency.	Observations.
	fathoms.	fathoms.			
Palmipes, *Link.*					
membranaceus, *Retz.*					
Asterina, *Nardo.*					
gibbosa, *Penn.*					
Goniaster, *Agassiz.*					
Templetoni, *Thomps.*					
equestris, *Gmelin.*					
Abbensis, *Forbes*					
Asterias, *Linnæus.*					
aurantiaca, *Linn.*					
Luidia, *Forbes.*					
fragilissima, *Forbes*					
Savignii, *Audouin*					
Order ECHINOIDEA.					
Fam. **Cidaridæ.**					
Cidaris, *Leske.*					
papillata, *Leske.*					
Echinus, *Linnæus.*					
sphæra, *Müller.*					
Flemingii, *Ball*					
miliaris, *Leske*					
lividus, *Lamk.*					
melo, *Lamk.*					
Norvegicus, *D. & K.*					
neglectus, *Lamk.*					
Fam. **Clypeasteridæ.**					
Echinocyamus, *Leske.*					
pusillus, *Müller*					
Echinarachnius, *Leske.*					
placenta, *Gmelin*					
Fam. **Spatangidæ.**					
Spatangus, *Klein.*					
purpureus, *Müller*					
Brissus, *Klein.*					
lyrifer, *Forbes*					
Amphidotus, *Agassiz.*					
cordatus, *Penn.*					
roseus, *Forbes*					
gibbosus, *Barrett*					
Order HOLOTHUROIDEA.					
Fam. **Pentactidæ.**					
Cucumaria, *Cuvier.*					
frondosa, *Gunner*					

48

ECHINODERMATA.

Species.	Range.	Found living at	Ground.	Frequency.	Observations.
	fathoms.	fathoms.			
CUCUMARIA, *Cuvier.*					
? fucicola, *Forbes & Goodsir*					
pentactes, *Müller*					
? Montagui, *Flem.*					
? Neillii, *Flem.*					
? dissimilis, *Flem.*					
fusiformis, *Forbes & Goodsir* ..					
Hyndmanni, *Thomps.*					
OCNUS, *Forbes* (= CUCUMARIA ?).					
brunneus, *Forbes*					
lacteus, *Forbes & Goodsir*					
PSOLINUS, *Forbes.*					
brevis, *Forbes & Goodsir*					
Fam. **Thyonidæ.**					
THYONE, *Oken.*					
fusus, *Müll.* (papillosa, *Abildg.*)					
raphanus, *Duben & Koren*					
communis, *F. & G.* (THYONIDIUM,					
D. & K.)					
Portlockii, *Forbes*					
Drummondii, *Thomps.*					
pellucida, *Vahl* (CUCUMARIA hya-					
lina, *F.*)					
HOLOTHURIA, *Linn.*					
nigra, *Couch*					
intestinalis					
tubulosa, *Linn.*					
Fam. **Psolidæ.**					
PSOLUS, *Oken.*					
phantopus, *Linn.*					
Forbesii					
Fam. **Synaptidæ.**					
SYNAPTA, *Esch.*					
inhærens, *Müll.*					
digitata, *Mont.*					

49

G 2

POLYZOA.

Species.	Range.	Found living at	Ground.	Frequency.	Observations.
Order I. INFUNDIBULATA.	fathoms.	fathoms.			
Suborder I. Cheilostomata.					
Fam. II. **Salicornariadæ.**					
Salicornaria, *Cuvier.*					
farciminoides, *Johnst.*					
Johnstoni, *Busk*					
sinuosa, *Hassall*					
Onchopora, *Busk.*					
borealis, *Busk.*					
Fam. III. **Cellulariadæ.**					
Cellularia, *Pallas.*					
Peachii, *Busk*					
cuspidata, *Busk*					
Menipea, *Lamouroux.*					
ternata, *Ellis & Soland.*					
Scrupocellaria, *Van Beneden.*					
scrupea, *Busk*					
scruposa, *Linn.*					
Canda, *Lamouroux.*					
reptans, *Pall.*					
Fam. IV. **Scrupariadæ.**					
Scruparia, *Oken.*					
chelata, *Linn.*					
clavata, *Hincks*					
Salpingia, *Coppin.*					
Hassallii. *Coppin.*					
Hippothoa, *Lamouroux.*					
catenularia, *Jameson*					
divaricata, *Lamx.*					
Ætea, *Lamouroux.*					
anguina. *Linn.*					
truncata. *Landsb.*					
recta, *Hincks*					
Beania, *Johnston.*					
mirabilis, *Johnst.*					
Fam. VI. **Gemellariadæ.**					
Gemellaria, *Savigny.*					
loricata, *Linn.*					

50

Species.	Range.	Found living at	Ground.	Frequency.	Observations.
	fathoms.	fathoms.			
NOTAMIA, *Fleming.*					
bursaria, *Linn*............					
Fam. VII. Cabereadæ.					
CABEREA, *Lamouroux.*					
Hookeri, *Flem*............					
Boryi, *Aud*..............					
Fam. VIII. Bicellariadæ.					
BICELLARIA, *De Blainville.*					
ciliata, *Linn*.					
Alderi, *Busk*					
BUGULA, *Oken.*					
neritina, *Linn*............					
flabellata, *J. V. Thomps*......					
avicularia, *Pall*.					
plumosa, *Pall*.					
Murrayina, *Bean*..........					
turbinata, *Alder*					
fastigiata, *Fabr*.					
Fam. IX. Flustradæ.					
FLUSTRA, *Linn.*					
foliacea, *Linn*.............					
papyracea, *Ellis*					
truncata, *Linn*.					
Barleei, *Busk*					
CARBASEA, *Gray.*					
papyrea, *Pall*.					
Fam. X. Membraniporidæ.					
MEMBRANIPORA, *De Blainville.*					
membranacea, *Linn*.					
pilosa, *Pall*...............					
coriacea, *Esper*					
lineata, *Linn*.					
Flemingii, *Busk*					
Rosselii, *Audouin*					
Lacroixii, *Savigny*					
monostachys, *Busk*					
hexagona, *Busk*					
Pouilletii, *Audouin*					
spinifera, *Johnst*...........					
craticula, *Alder*					
unicornis, *Flem*.					
imbellis, *Hincks*					

51

POLYZOA.

Species.	Range.	Found living at	Ground	Fre-quency.	Observations.
	fathoms	fathoms			
LEPRALIA, *Johnston.*					
Brongniartii, *Aud.*					
Landsborovii, *Johnst.*					
reticulata, *Macgillivray*					
auriculata, *Hassall*					
concinna, *Busk*					
verrucosa, *Esper*					
violacea, *Johnst.*					
spinifera, *Johnst.*					
trispinosa, *Johnst.*					
coccinea, *Abildg.*					
linearis, *Hassall*					
ciliata, *Pall.*					
Gattyæ, *Landsb.*					
Hyndmanni, *Johnst.*					
variolosa, *Johnst.*					
nitida, *Fabr.*					
annulata, *Fabr.*					
bispinosa, *Johnst.*					
Peachii, *Johnst.*					
ventricosa, *Hassall*					
melolontha, *Landsb.*					
innominata, *Couch*					
punctata, *Hassall*					
figularis, *Johnst.*					
pertusa, *Esper*					
Pallasiana					
labrosa, *Busk*					
simplex, *Johnst.*					
Malusii, *Aud.*					
granifera, *Johnst.*					
hyalina, *Linn.*					
ansata, *Johnst.*					
unicornis, *Flem.*					
ringens, *Busk*					
fissa, *Busk*					
Cecilii, *Audouin*					
Barleei, *Busk*					
canthariformis, *Busk*					
umbonata, *Busk*					
discoidea, *Busk*					
bella, *Busk*					
monodon, *Busk*					
alba, *Hincks*					
eximia, *Hincks.*					
Woodiana, *Busk*					
ALYSIDOTA, *Busk.*					
Alderi, *Busk*					

Species.	Range.	Found living at	Ground.	Fre-quency.	Observations.
	fathoms.	fathoms.			
Fam. XI. Celleporidæ.					
CELLEPORA, *Fabr.*					
pumicosa, *Linn.*					
Hassallii, *Johnst.* .... , ...					
vitrina, *Couch*					
ramulosa. *Linn.*					
Skenei, *Ellis & Soland.*					
tubigera, *Busk*					
armata, *Hincks*					
avicularis, *Hincks*					
Fam. XII. Escharidæ.					
ESCHARA, *Ray.*					
foliacea, *Ellis & Soland.*					
cervicornis, *Soland.*					
cribraria. *Johnst.*					
RETEPORA, *Lamarck.*					
cellulosa, *Linn.*					
Beaniana, *King*					
Suborder II. CYCLOSTOMATA.					
Fam. I. Tubuliporidæ.					
TUBULIPORA, *Lamarck.*					
patina, *Linn.*					
hispida, *Flem.*					
penicillata, *Johnst.*					
truncata, *Jameson*					
lobulata, *Hassall*					
phalangea, *Couch*					
flabellaris, *Fabr.*					
serpens, *Linn.*					
hyalina, *Couch*					
DIASTOPORA, *Lamouroux.*					
obelia, *Flem.*					
IDMONEA, *Lamouroux.*					
Atlantica, *Forbes*					
PUSTULIPORA, *De Blainville.*					
proboscidea, *M.-Edw.*					
deflexa, *Couch*					
Oreadensis, *Busk*					
ALECTO, *Lamouroux.*					
granulata, *M.-Edw.*					
major, *Johnst.*					
dilatans, *Johnst.*					
incurvata, *Hincks*					

53

Species.	Range.	Found living at	Ground.	Frequency.	Observations.
	fathoms.	fathoms.			

Fam. 11. **Crisiadæ.**

CRISIA, *Lamouroux.*
eburnea, *Linn.* ....
denticulata, *Lamk.* ....
aculeata, *Hassall* ....
geniculata, *M.-Edw.* ....
CRISIDIA, *M.-Edwards.*
cornuta, *Linn.* ....
setacea, *Couch* ....

Suborder III. CTENOSTOMATA.

Fam. 1. **Alcyonidiadæ.**

ALCYONIDIUM, *Lamouroux.*
gelatinosum, *Pallas* ....
hirsutum, *Flem.* ....
parasiticum, *Flem.* ....
mamillatum, *Alder* ....
albidum, *Alder* ....
hexagonum, *Hincks* ....
CYCLUM, *Hass.*
papillosum, *Hass.* ....
SARCOCHITUM, *Hass.*
polyoum, *Johnst.* ....

Fam. II. **Vesicalariadæ.**

AMATHIA, *Lamouroux.*
lendigera, *Linn.* ....
VESICULARIA, *Thompson.*
spinosa, *Linn.* ....
VALKERIA, *Fleming.*
cuscuta, *Ellis* ....
uva, *Linn.* ....
pustulosa, *Johnst.* ....
MIMOSELLA, *Hincks.*
gracilis, *Hincks* ....
AVENELLA, *Dalyell.*
fusca, *Dalyell* ....
NOTELLA, *Gosse.*
stipata, *Gosse* ....
BOWERBANKIA, *Farre.*
imbricata, *Johnst.* ....
FARRELLA, *Ehrenberg.*
repens, *Johnst.* ....
elongata ....
gigantea ....
54

Species.	Range.	Found living at	Ground.	Fre-quency.	Observations.
	fathoms.	fathoms.			
FARRELLA, *Ehrenberg.*					
pedicellata, *Alder*					
dilatata, *Hincks*					
ANGUINELLA, *Van Beneden.*					
palmata, *V. Ben.*					
BUSKIA, *Alder.*					
nitens, *Alder*					
Fam. III. **Pedicellinidæ.**					
PEDICELLINA, *Sars.*					
echinata, *Sars*					
Belgica					
gracilis					

·

Sub-Kingdom CŒLENTERATA.

CLASS HYDROZOA.

⁎ This list is compiled from Dr. Johnston's "British Zoophytes" (2nd edit.), Forbes's "British Naked-eyed Medusæ," and the works of Mr. Alder, Prof. Allman, Mr. Cobbold, Mr. Gosse, Prof. Greene, Rev. Thomas Hincks, Prof. Huxley, Dr. T. Strethill Wright, &c.

Order CORYNIDÆ.

Fam. I. **Coryniadæ.**

CLAVA, *Gmelin.*
multicornis, *Johnston* (repens, *T. S. Wright*; discreta, *Allman*) ....
cornea, *T. S. Wright* ....
membranacea, *T. S. Wright* ..
VORTICLAVA, *Alder.*
humilis, *Alder* ....
LAR, *Gosse.*
Sabellarum, *Gosse* ....
HYDRACTINIA, *Van Beneden* (Po-docoryna. *Sars*).
echinata, *Flem.* ....
carnea, *Sars* ....
MYRIOTHELA, *Sars.*
arctica, *Sars* ....

55

H

Species.	Range.	Found living at	Ground.	Fre-quency.	Observations.
	fathoms.	fathoms.			
CLAVATELLA, *Hincks.*					
prolifera, *Hincks*					
CORYNE, *Gaërtner.*					
pusilla, *Ehr.*					
Sarsii, *Lovén* (decipiens, *Dujar-din*) ....,					
ramosa, *Ehr.* (Listerii, *Van Be-neden*)					
sessilis, *Gosse*					
gravata, *T. S. Wright*					
eximia, *Allman* ....,					
implexa. *T. S. Wright* (TUBU-LARIA implexa, *Alder*; ? C. Briareus, *Allman*)					
Cerberus, *Gosse*					
stauridia, *Dujardin*. [Should be referred to the next genus.]..					
STAURIDIA, *T. S. Wright.*					
producta, *T. S. Wright*					
TRICHYDRA, *T. S. Wright.*					
pudica, *T. S. Wright*					
Fam. II. Tubulariadæ.					
EUDENDRIUM, *Ehrenberg.*					
ramosum, *Linn.*					
rameum, *Pall.*					
capillare, *Alder* ....:...					
arbuscula, *T. S. Wright*					
ATRACTYLIS, *T. S. Wright.*					
ramosa, *Van Beneden*					
repens, *T. S. Wright*					
sessilis, *T. S. Wright*					
DICORYNE, *Allman.*					
conferta. *Alder* (EUDEND. con-fertum, *Alder*)					
GARVEIA, *T. S. Wright.*					
nutans, *T. S. Wright*					
BIMERIA, *T. S. Wright.*					
vestita (MANICELLA fusca, *All-man*)....................					
TUBULARIA, *Linnæus.*					
indivisa, *Linn.*					
Dumortierii, *Van Beneden*					
larynx, *Ellis*................					
gracilis, *Harvey*					
CORYMORPHA, *Sars.*					
nutans. *Sars*................,					
nana, *Alder*					

56

Species.	Range.	Found living at	Ground.	Fre- quency.	Observations.
	fathoms.	fathoms.			

Order SERTULARIDÆ.

Fam. I. Sertulariadæ.

HALECIUM, *Oken.*
  halecinum, *Ellis* ....
  Beanii, *Johnst.* ....
  muricatum, *Ellis & Soland.* .....
  labrosum, *Alder* ....
  tenellum, *Hincks* ....
SERTULARIA, *Linnæus.*
  polyzonias, *Linn.* ....
  tricuspidata, *Alder* ....
  tenella, *Alder* ....
  Gayi, *Lamx.* ....
  rugosa, *Ellis.* ....
  rosacea, *Linn.* ....
  pumila, *Linn.* ....
  gracilis, *Hassall* ....
  Evansii, *Ellis & Soland.* ......
  nigra, *Pallas* ....
  pinnata, *Pallas.* ....
  alata, *Hincks* ....
  pinaster, *Ellis & Soland.* ......
  Margareta, *Hassall* ....
  fallax, *Johnst.* ....
  tamarisca, *Linn.* ....
  abietina, *Linn.* ....
  filicula, *Ellis & Soland.* ......
  operculata, *Linn.* ....
  argentea, *Ellis & Soland.* ....
  cupressina, *Linn.* ....
  fusca, *Johnst.* ....
THUIARIA, *Fleming.* .
  thuia, *Linn.* ....
  articulata, *Pallas* ....
ANTENNULARIA, *Lamarck.*
  antennina, *Linn.* ....
  ramosa, *Lamx.* ....
PLUMULARIA, *Lamarck.*
  falcata, *Linn.* ....
  cristata, *Lamk.* ....
  pennatula, *Ellis & Soland.* ....
  myriophyllum, *Linn.* ....
  tubulifera, *Hincks* ....
  pinnata, *Linn.* ....
  setacea, *Ellis* ....
  Catherina, *Johnst.* ....
  echinulata, *Lamk.* ....

57

Species.	Range.	Found living at	Ground.	Frequency.	Observations.
	fathoms.	fathoms.			
PLUMULARIA, *Lamarck*.					
similis, *Hincks*					
frutescens, *Ellis & Soland.*					
halecioides, *Alder*					
obliqua, *Saunders* (LAOMEDEA					
obliqua, *Johnst.*)					
Fam. II. **Campanulariadæ.**					
LAOMEDEA, *Lamouroux*.					
dichotoma, *Linn.*					
longissima, *Pallas*					
geniculata, *Linn.*					
flexuosa, *Hincks*					
Lovéni, *Allman*					
gelatinosa, *Pallas*					
angulata, *Hincks*					
neglecta, *Alder*					
pulchella, *Wyville Thomson* . .					
lacerata, *Johnst.*					
tenuis, *Allman*					
acuminata, *Alder*					
CAMPANULARIA, *Lamarck*.					
volubilis, *Linn.*					
Johnstoni, *Alder*					
Hincksii, *Alder*					
raridentata, *Alder*					
integra, *Macgillivray*					
caliculata, *Hincks*					
verticillata, *Linn.*					
[intertexta, *Couch* — a very					
doubtful species]					
CALICELLA, *Hincks*.					
dumosa, *Flem.*					
gracillima, *Alder*					
parvula, *Hincks*					
syringa, *Linn.*					
fastigiata, *Alder*					
humilis, *Hincks*					
RETICULARIA, *Wyville Thomson*.					
serpens, *Hassall*					
GRAMMARIA, *Stimpson*.					
ramosa, *Alder*					
COPPINIA, *Hassall*. [The position					
of this genus is doubtful.]					
arcta, *Dalyell*					

Species.	Range.	Found living at.	Ground.	Frequency.	Observations.
	fathoms.	fathoms.			
Order CALYCOPHORID.E.					
Fam. **Diphydæ.**					
DIPHYES, *Cuvier.*					
appendiculata, *Eschscholtz*				'	
Order PHYSOPHORID.E.					
Fam. I. **Stephanomiadæ.**					
(?) HALISTEMMA, *Huxley.*					
rubrum, *Vogt*					
Fam. II. **Physaliadæ.**					
PHYSALIA, *Lamarck.*					
pelagica, *Eschscholtz*					
VELELLA, *Lamarck.*					
spirans, *Forsk.*					
Order MEDUSID.E.					
Fam. I. **Willsiadæ.**					
WILLSIA, *Forbes.*					
stellata, *Forbes*..............					
Fam. II. **Oceanidæ.**					
TURRIS, *Lesson.*					
digitalis, *O. F. Müller*					
neglecta, *Lesson*					
constricta, *Patterson*..........				'	
SAPHENIA, *Eschscholtz.*					
dinema, *Péron*				i	
Titania, *Gosse*				'	
OCEANIA, *Péron.*					
octona, *Flem.*					
episcopalis, *Forbes*				i	
turrita, *Forbes*				i	
globulosa, *Forbes*					
ducalis, *Forbes & Goodsir*......					
pusilla, *Gosse*					
Fam. III. **Æquoreadæ.**					
STOMOBRACHIUM, *Brandt.*					
octocostatum, *Sars*					
POLYXENIA, *Eschscholtz.*					
Alderi, *Forbes*					

59

Species.	Range.	Found living at	Ground.	Fre-quency.	Observations
	fathoms.	fathoms.			
.Equorea, *Péron.*					
Forskalii, *Forbes*					
Forbesiana, *Gosse*...........					
vitrina, *Gosse* ......•.......					
formosa, *Greene*					
sp., *Greene*					
Fam. IV. Circeadæ.					
Circe, *Mertens.*					
rosea, *Forbes* ....|					
Fam. V. Geryoniadæ.					
Geryonia, *Péron.*					
appendiculata, *Forbes*					
Tima, *Eschscholtz.*					
Bairdii, *Johnst.*..............					
Geryonopsis, *Forbes.*					
delicatula, *Forbes*............					
Tiaropsis, *Agassiz.*					
Pattersonii, *Greene*					
Thaumantias, *Eschscholtz.*					
pilosella, *Forbes*					
quadrata, *Forbes*					
æronautica, *Forbes* ....'					
octona, *Forbes*					
maculata, *Forbes*					
melanops, *Forbes*					
globosa, *Forbes*.............					
convexa, *Forbes*					
gibbosa, *Forbes*					
lineata, *Forbes*.............					
pileata, *Forbes*.............					
Sarnica, *Forbes*.............					
Thompsoni, *Forbes*					
hemisphærica, *O. F. Müller*....					
inconspicua, *Forbes* ....|					
punctata, *Forbes* ....'					
lucifera, *Forbes*					
Buskiana, *Gosse*					
corynetes, *Gosse*					
undulata, *Forbes & Goodsir*....					
confluens, *Forbes & Goodsir*....					
achron, *Cobbold*					
neglecta, *Greene*					
typica, *Greene*					
Sladderia, *Forbes.*					
halterata, *Forbes*					
catenata, *Forbes & Goodsir*					

Species.	Range.	Found living at	Ground.	Frequency.	Observations.
	fathoms.	fathoms.			
Fam. VI. **Sarsiadæ.**					
PLANCIA, *Forbes & Goodsir.*					
gracilis, *Forbes & Goodsir*					
GOODSIREA, *Strethill Wright.*					
mirabilis, *S. Wright*					
SARSIA, *Lesson.*					
tubulosa, *Sars*					
pulchella, *Forbes*					
gemmifera, *Forbes*					
prolifera, *Forbes*					
HIPPOCRENE, *Mertens.*					
Britannica, *Forbes*					
nigritella, *Forbes*					
pyramidata, *Forbes & Goodsir* . .					
crucifera, *Forbes & Goodsir*					
simplex, *Forbes & Goodsir*					
dinema, *Greene*					
LIZZIA, *Forbes.*					
octopunctata, *Sars*					
blondina, *Forbes*					
sp., *Claparède*					
MODEERIA, *Forbes.*					
formosa, *Forbes*					
DIPLONEMA, *Greene.*					
Islandica, *Greene*					
EUPHYSA, *Forbes.*					
aurata, *Forbes*					
STEENSTRUPIA, *Forbes.*					
rubra, *Forbes*					
flaveola, *Forbes*					
Owenii, *Greene*					
Order LUCERNARIDÆ.					
Fam. I. **Lucernariadæ.**					
LUCERNARIA, *Müller.*					
auricula, *Fabr.*					
campanulata, *Lamx.*					
fascicularis, *Flem.*					
DEPASTRUM, *Gosse.*					
stellifrons, *Gosse*					
CARDUELLA, *Allman.*					
cyathiformis, *Sars*					
Fam. II. **Pelagidæ.**					
AURELIA, *Péron.*					
aurita, *O. F. Müller*					
campanula, *O. Fabricius*					
61					

Species.	Range.	Found living at	Ground.	Frequency.	Observations
CYANEA, *Péron.*	fathoms.	fathoms.			
capillata, *Linn.*					
Lamarckii, *Péron*					
PELAGIA, *Péron et Lesueur.*					
cyanella, *Péron et Lesueur*					
CHRYSAORA, *Péron.*					
hysoscella, *Linn.*					
Fam. III. Rhizostomidæ.					
CASSIOPEIA, *Péron.*					
lunulata, *Flem.*					
RHIZOSTOMA, *Cuvier.*					
pulmo, *Gmel.*					

CLASS ACTINOZOA.

*** The list of Zoantharia is taken from Gosse's "Actinologia."

Order ZOANTHARIA.

Fam. I. Actiniadæ.

ACTINOLOBIA, *Blainville.*					
dianthus, *Blainv.*					
SAGARTIA, *Gosse.*					
bellis, *Ellis*					
miniata, *Gosse*					
rosea, *Gosse*					
ornata, *Holdsworth.*					
ichthystoma, *Gosse*					
nivea, *Gosse*					
sphyrodeta, *Gosse*					
pallida, *Holdsworth*					
pura, *Alder*					
coccinea, *Müll.*					
troglodytes, *Johnst.*					
viduata, *Müll.*					
parasitica, *Johnst.*					
chrysosplenium, *Cocks*					
ADAMSIA, *Forbes.*					
palliata, *Forbes*					
PHELLIA, *Gosse.*					
murociucta, *Gosse*					
gausapata, *Gosse*					
picta, *Gosse*					
Brodricii, *Gosse*					
GREGORIA, *Gosse.*					
fenestrata, *Gosse*					
AIPTASIA, *Cocks.*					
Couchii, *Cocks*					

62

ACTINOZOA.

Species.	Range.	Found living at	Ground.	Fre-quency.	Observations.
	fathoms.	fathoms.			
ANTHEA, *Gaertner.*					
cereus, *Ellis*					
ACTINIA, *Linn.*					
mesembryanthemum, *Ellis*					
BOLOCERA, *Johnst.*					
Tuediæ, *Johnst.*					
eques, *Gosse*					
BUNODES, *Gosse.*					
gemmacea, *Ellis*					
thallia, *Gosse*					
Ballii, *Cocks*					
coronata, *Gosse*					
TEALIA, *Gosse.*					
digitata, *Müll.*					
tuberculata, *Cocks*					
crassicornis, *Müll.*					
HORMATHIA, *Gosse.*					
margarita, *Gosse*					
STOMPHIA, *Gosse.*					
Churchiæ, *Gosse*					
ILYANTHUS, *Forbes.*					
Scoticus, *Forbes*					
Mitchellii, *Gosse*					
PEACHIA, *Gosse.*					
hastata, *Gosse*					
undata, *Gosse*					
triphylla, *Gosse*					
cylindrica, *Reid*					
HALCAMPA, *Gosse.*					
chrysanthemum, *Peach*					
microps, *Gosse*					
EDWARDSIA, *Quatrefages.*					
callimorpha, *Gosse*					
carnea, *Gosse*					
Beautempsii, *Quatref.*					
ARACHNACTIS, *Sars.*					
albida, *Sars*					
CERIANTHUS, *J. Haime.*					
Lloydii, *Gosse*					
vermicularis, *Forbes*					
CAPNEA, *Forbes.*					
sanguinea, *Forbes*					
AURELIANIA, *Gosse.*					
angusta, *Gosse*					
heterocera, *Thomps.*					
CORYNACTIS, *Allman.*					
viridis, *Allman*					

Fam. II. **Zoanthidæ.**

ZOANTHUS, *Cuvier.*
Couchii, *Johnst.* ....

G3

Species.	Range.	Found living at	Ground.	Frequency.	Observations.
	fathoms.	fathoms.			

ZOANTHUS, *Cuvier.*
  sulcatus, *Gosse* ....
  Alderi, *Gosse* ....

Fam. III. Caryophylleadæ.

CYATHINA, *Ehrenberg.*
  Smithii, *Flem.* ....
PARACYATHUS, *M.-Edwards.*
  Taxilianus, *Gosse* ....
  Thulensis, *Gosse* ....
  pteropus, *Gosse* ....
DESMOPHYLLUM, *Ehrenberg.*
  Stokesii, *M.-Edw.* ....
SPHENOTROCHUS, *M.-Edwards.*
  Macandrewanus, *M.-Edw.* ....
  Wrightii, *Gosse* ....
ULOCYATHUS, *Sars.*
  arcticus, *Sars* ....
OCULINA, *Lamarck.*
  prolifera, *Linn.* ....
HOPLANGIA, *Gosse.*
  Durotrix, *Gosse* ....
BALANOPHYLLIA, *Wood.*
  regia, *Gosse* ....

Order ALCYONARIA.

Fam. I. Pennatuladæ.

PENNATULA, *Linnæus.*
  phosphorea, *Linn.* .....
VIRGULARIA, *Lamarck.*
  mirabilis, *Linn.* ....
PAVONARIA, *Cuvier.*
  quadrangularis, *Pall.* ....

Fam. II. Alcyonidæ.

ALCYONIUM, *Linnæus.*
  digitatum, *Linn.* ....
  glomeratum ....
SARCODICTYON, *Forbes.*
  catenata, *Forbes* ....
  agglomerata ....

Fam. III. Gorgoniadæ.

GORGONIA, *Linnæus.*
  verrucosa, *Linn.* ....
  pinnata, *Linn.* ....
  anceps, *Ellis* ....
PRIMNOA, *Lamarck.*
  lepadifera, *Linn.* ....

64

ACTINOZOA.

Species.	Range.	Found living at	Ground	Frequency.	Observations.
	fathoms	fathoms.			

Order CTENOPHORA.

  Fam. I. **Cydippidæ.**

Cydippe, *Esch.*
pileus, *Müll.* . . . .
Flemingii, *Forbes.* . . . .
infundibulum, *Müll.* . . . .
lagena, *Forbes* . . . .
pomiformis, *Paterson* . . . .

  Fam. II. **Calymnidæ.**

Bolina, *Paterson.*
Hibernica, *Paterson* . . . . . .

  Fam. III. **Beroidæ.**

Beroë, *Müller.*
cucumis, *Fabr.* . . . .
fulgens, *Flem.* . . . .
borealis, *Less.* . . . .
Alcinoë, *Cuv.*
rotunda . . . .
Smithii . . . .

Sub-Kingdom Protozoa.

FORAMINIFERA.

₊ This list of British Foraminifera is taken from Prof. Williamson's "Recent Foraminifera of Great Britain." published by the Ray Society.

Species.	Range.	Found living at	Ground.	Frequency.	Observations.
	fathoms.	fathoms.			
Proteonina, *Williamson.*					
fusiformis, *Williamson*					
pseudospiralis, *Williamson*					
Orbulina, *D'Orbigny.*					
universa, *D'Orb.*					
Lagena, *Walker.*					
vulgaris, *Williamson*					
var. clavata					
var. perlucida					
var. semistriata					
var. striata					
var. interrupta...........					
var. gracilis					
var. substriata					
Entosolenia, *Ehrenberg.*					
globosa, *Walker*					
var. lineata					
costata, *Williamson*..........					
marginata, *Walker*					
var. lucida					
var. ornata					
var. lagenoides					
var. quadrata					
squamosa, *Mont.*					
var. scalariformis					
var. catenulata...........					
var. hexagona					
Lingulina, *D'Orbigny.*					
carinata, *D'Orb.*.............					
Nodosaria, *Lamarck.*					
radicula, *Linn.*.............					
pyrula, *D'Orb.*.............					
Dentalina, *D'Orbigny.*					
subarcuata, *Mont.*...........					
var. jugosa					
legumen, *Linn.*.............					
var. linearis					
Frondicularia, *Defrance.*					
spathulata, *Williamson*........					
Archiaeiana, *D'Orb.*..........					
Cristellaria, *Lamarck.*					
calcar, *Linn.*...............					
var. rotifera.............					
var. oblonga.............					

Species.	Range.	Found living at	Ground.	Frequency	Observations.
	fathoms.	fathoms.			
CRISTELLARIA. *Lamarck.*					
subarcuatula. *Walker*					
var. costata					
NONIONINA. *D'Orbigny.*					
Barleeana, *Williamson*........					
crassula, *Walker*					
Jeffreysii, *Williamson*					
elegans, *Williamson*					
NUMMULINA. *D'Orbigny.*					
planulata. *Lam*...............					
POLYSTOMELLA. *Lamarck.*					
crispa, *Linn*.................					
umbilicatula, *Walker*					
var. incerta					
PENEROPLIS. *Montfort.*					
planatus. *Ficht. & Moll*					
PATELLINA. *Williamson.*					
corrugata, *Williamson*					
ROTALINA. *D'Orbigny.*					
Beccarii, *Linn*........					
inflata, *Mont*................					
turgida, *Williamson*..........					
oblonga, *Williamson*					
concamerata. *Mont.*					
nitida, *Williamson*					
mamilla, *Williamson*					
ochracea, *Williamson*					
fusca, *Williamson*............					
GLOBIGERINA, *D'Orbigny.*					
bulloides, *D'Orb.*					
PLANORBULINA, *D'Orbigny.*					
vulgaris, *D'Orb.*					
TRUNCATULINA, *D'Orbigny.*					
lobatula, *Walker*					
BULIMINA, *D'Orbigny.*					
pupoides, *D'Orb.*					
var. marginata...........					
var. spinulosa					
var. fusiformis					
var. compressa					
var. convoluta					
elegantissima, *D'Orb.*					
scabra?, *Williamson*..........					
UVIGERINA, *D'Orbigny.*					
pygmæa, *D'Orb.*					
angulosa, *Williamson*					
CASSIDULINA, *D'Orbigny.*					
lævigata, *D'Orb.*					
obtusa, *Williamson*					

67

Species.	Range.	Found living at	Ground.	Frequency.	Observations.
	fathoms	fathoms			
POLYMORPHINA, *D'Orbigny.*					
lactea, *Walker*					
var. acuminata					
var. oblonga					
var. fistulosa					
var. concava					
var. communis					
myristiformis, *Williamson*					
TEXTULARIA, *Defrance.*					
cuneiformis, *D'Orb.*					
var. conica					
variabilis, *Williamson*					
var. spathulata					
var. difformis					
var. lævigata					
BILOCULINA, *D'Orbigny.*					
ringens, *D'Orb*					
var. carinata					
SPIROLOCULINA, *D'Orbigny.*					
depressa, *D'Orb.*					
var. rotundata					
var. cymbium					
MILIOLINA, *Williamson.*					
trigonula, *Lamk.*					
seminulum, *Linn.*					
var. oblonga					
var. disciformis					
bicornis, *Walker*					
var. elegans					
var. angulata					
VERTEBRALINA, *D'Orbigny.*					
striata, *D'Orb.*					
SPIRILLINA. *Ehrenberg.*					
foliacea, *Philippi*					
perforata, *Schultze*					
arenacea, *Williamson*					
margaritifera, *Williamson*					

LIST OF BRITISH SPONGES.

TETHYA, *Lamarck.*			
cranium, *Johnst.* (not *Müll.*)			
lyncurium, *Johnst.*			
GEODIA, *Lamarck.*			
Zetlandica, *Johnst.*			

68

Species.	Range.	Found living at	Ground.	Fre-quency.	Observations.
	fathoms.	fath oms.			
PACHYMATISMA, *Bowerbank*.					
Johnstonia, *Bowb.* (HALICHON-					
DRIA, *Johnst.*)					
HALICHONDRIA, *Fleming.*					
panicea, *Johnst.*					
coalita, *Johnst.*					
coccinea, *Bowb. MS.*					
glabra, *Bowb. MS.*					
inconspicua, *Bowb. MS.*					
caduca, *Bowb. MS.*					
distorta, *Bowb. MS.*					
Dickiei, *Bowb. MS.*					
Batei, *Bowb. MS.*					
lingua. *Bowb. MS.*					
corrugata, *Bowb. MS.*					
granulata, *Bowb. MS.*					
Thompsoni, *Bowb. MS.*					
plumosa, *Johnst.*					
incrustans, *Johnst.*					
fallax, *Bowb. MS.*					
paupera, *Bowb. MS.*					
Clarkei, *Bowb. MS.*					
variantia, *Bowb. MS.*					
Hyndmanni, *Bowb. MS.*					
nigricans, *Bowb. MS.*					
Ingalli, *Bowb. MS.*					
fimbriata, *Bowb. MS.*					
albula, *Bowb. MS.*					
farinaria, *Bowb. MS.*					
HYMENIACIDON, *Bowerbank* (HA-					
LICHONDRIA, *Johnst.*).					
Thomasii, *Bowb. MS.*					
fragilis, *Bowb. MS.*					
Brettii, *Bowb. MS.*					
albescens, *Johnst.*					
caruncula, *Bowb. MS.*					
Alderi, *Bowb. MS.*					
perlevis, *Johnst.*					
aurea, *Johnst.*					
pachyderma, *Bowb. MS.*					
crustula, *Bowb. MS.*					
sanguinea, *Johnst.*					
armatum, *Bowb. MS.*					
floreum, *Bowb. MS.*					
carnosa, *Bowb. MS.*					
viridans, *Bowb. MS.*					
sulphurea, *Bowb. MS.*					
clavigera, *Bowb. MS.*					
subclavata, *Bowb. MS.*					
lactea, *Bowb. MS.*					
Dujardinii (HALISARCA), *Johnst.*					

Species.	Range.	Found living at	Ground.	Frequency.	Observations.
	fathoms.	fathoms			
HYMENIACIDON, *Bowerbank* (HA-LICHONDRIA, *Johnst.*).					
celata, *Johnst.*					
HALINA, *Bowerbank* (HALICHON-DRIA, *Johnst.*).					
suberea (HALICHONDRIA), *Johnst.*					
ficus, *Johnst.*					
carnosa, *Johnst.*					
Bucklandi, *Bowb. MS.*					
ISODICTYA, *Bowerbank* (HALICHON-DRIA, *Johnst.*).					
Peachii, *Bowb. MS.*					
rosea, *Bowb. MS.*					
permollis, *Bowb. MS.*					
indistincta, *Bowb. MS.*					
indefinita, *Bowb. MS.*					
Macandrewii, *Bowb. MS.*					
dichotoma, *Bowb. MS.*					
cinerea, *Johnst.*					
ramusculus, *Bowb. MS.*					
simulo, *Bowb. MS.*					
mammeata, *Bowb. MS.*					
fucorum, *Johnst.*					
Alderi, *Bowb. MS.*					
Normani, *Bowb. MS.*					
lobata, *Johnst.*					
Barleei, *Bowb. MS.*					
gracilis, *Bowb. MS.*					
Gregorii, *Bowb. MS.*					
Beanii, *Bowb. MS.*					
clava, *Bowb. MS.*					
infundibuliformis, *Johnst.*					
DESMACIDON, *Bowerbank* (HALI-CHONDRIA, *Johnst.*).					
ægagropila, *Johnst.*					
fruticosa, *Johnst.*					
RAPHYRUS, *Bowerbank.*					
Griffithsii, *Bowb. MS.*					
DICTYOCYLINDRUS, *Bowerbank* (HA-LICHONDRIA, *Johnst.*).					
stuposus, *Johnst.*					
Howsei, *Bowb. MS.*					
ramosus, *Johnst.*					
aculeatus, *Bowb. MS.*					
ventilabrum, *Bowb. MS.*					
fascicularis, *Bowb. MS.*					
rugosus, *Bowb. MS.*					
HALICLONA, *Bowerbank* (HALICHON-DRIA, *Johnst.*).					
palmata, *Johnst.*					
Montagui, *Johnst.*					

70

Species.	Range.	Found living at	Ground.	Frequency.	Observations.
	fathoms.	fathoms.			
HALICLONA, *Bowerbank* (HALICHON-DRIA, *Johnst.*).					
pygmæa, *Bowb.*					
serinta, *Johnst.*					
simulans, *Johnst.*					
columbæ. *Johnst.*					
gracilenta, *Bowb. MS.*					
MICROCIONA, *Bowerbank, MS.*					
atro-sanguinea, *Bowb. MS.*					
armata, *Bowb. MS.*					
carnosa, *Bowb. MS.*					
ambigua, *Bowb. MS.*					
lævis, *Bowb. MS.*					
spinulenta. *Bowb. MS.*					
HYMERAPHIA, *Bowerbank, MS.*					
vermiculata, *Bowb. MS.*					
stellifera, *Bowb. MS.*					
clavata, *Bowb. MS.*					
HYMEDESMIA, *Bowerbank, MS.*					
Zetlandica, *Bowb. MS.*					
HALYPHYSEMA, *Bowerbank, MS.*					
Tumanowiczii, *Bowb. MS.*					
EUPLECTELLA, *Owen.*					
mammillaris (HALICHONDRIA), *Johnst.*					
brevis, *Bowb. MS.*					
robusta, *Bowb. MS.*					
HALICNEMIA, *Bowerbank, MS.*					
patera, *Bowb. MS.*					
PHAKELLIA, *Bowb. MS.* (HALICHON-DRIA, *Johnst.*).					
ventilabrum, *Johnst.*					
DYSIDEA, *Johnston.*					
fragilis, *Johnst.*					
SPONGIA, *Linnæus.*					
pulchella, *Johnst.*					
limbata, *Johnst.*					
GRANTIA, *Fleming.*					
compressa, *Johnst.*					
ciliata, *Johnst.*					
ensata, *Bowb. MS.*					
tessellata, *Bowb. MS.*					
LEUCONIA, *Bowerbank* (GRANTIA, *Johnst.*).					
nivea, *Johnst.*					
fistulosa. *Johnst.*					
LEUCOSOLENIA, *Bowerbank, MS.* (GRANTIA, *Johnst.*).					
botryoides. *Johnst.*					
coriacea, *Johnst.*					
lacunosa, *Johnst.*					
contorta, *Bowb. MS.*					

71

www.ingramcontent.com/pod-product-compliance
Lightning Source LLC
Chambersburg PA
CBHW020246090426
42735CB00010B/1855